European Corrosion Congress (EUROCORR 2018)

Krakow, Poland
9-13 September 2018

ISBN: 978-1-5108-8637-7

Printed from e-media with permission by:

Curran Associates, Inc.
57 Morehouse Lane
Red Hook, NY 12571

CURRAN ASSOCIATES INC. proceedings.com

Some format issues inherent in the e-media version may also appear in this print version.

Printed by Curran Associates, Inc. (2019)

Polish Corrosion Society (PSK)
ul. Elblaska 133A
80-718 Gdansk, Poland

sekretarz@psk.org.pl

www.psk.org.pl

Additional copies of this publication are available from:

Curran Associates, Inc.
57 Morehouse Lane
Red Hook, NY 12571 USA
Phone: 845-758-0400
Fax: 845-758-2633
Email: curran@proceedings.com
Web: www.proceedings.com

TABLE OF CONTENTS

Posternumber	Paper title	Authors	Submission ID
Corrosion and Scale Inhibition (WP1)			
WP1.1	Influence of concentration and coating time on the barrier formation of sustainable N-acyl sarcosines to protect low carbon steel in aqueous solution.	Saad Kaskah (University of Koblenz-Landau, Koblenz) / Gitta Ehrenhaft (Institute of mechanics and material science, TH Mittelhessen University of Applied Sciences, Giessen) / Jörg Gollnick (Institute of mechanics and material science, TH Mittelhessen University of Applied Sciences, Giessen) / Christian Fischer (Department of Physics, University Koblenz-Landau, Koblenz)	105792
WP1.3	Effect of the Cerium (III) chloride on the corrosion inhibition of aluminium alloy	Marin Kurtela (Faculty of Mechanical Engineering and Naval Architecture , Zagreb, Croatia) / Vinko Šimunović (Faculty of Mechanical Engineering and Naval Architecture, Zagreb, Croatia) / Vesna Alar (Faculty of Mechanical Engineering and Naval Architecture , Zagreb, Croatia) / Suzana Jakovljević (Faculty of Mechanical Engineering and Naval Architecture , Zagreb, Croatia)	108162
WP1.4	Assessment of Glycine Propyl Ester Lauryl Hydrochloride as Sustainable Mild Steel Corrosion Inhibitor in 1M HCl	Saman Zehra (Aligarh Muslim University, Aligarh, India) / Mohammad Mobin (Aligarh Muslim University, Aligarh, India)	110337
WP1.5	Aqueous Corrosion through Galvanic Coupling in a Two-Phase Material: The Role of Interface Crystallography	Irshad khan (IIT BOMBAY, MUMBAI, India)	112527
WP1.6	Effect of water chemistry on scaling in heating systems	Georgeanna Kawaley (Mitsubishi Electric R&D Center Europe BV, Livingston West Lothian, United Kingdom)	113057
WP1.7	The Analysis for Phosphate-Free Coating Pretreatment Corrosion of Annealed Steel Sheet	Ting Shang (Shougang research institute of Technology, Beijing, China PR)	114317
WP1.8	Corrosion acceleration by RE3+ ions	Marco Oliveira (CICECO, DEMaC, University of Aveiro, Aveiro, Portugal) / Silvar Kallip (Institute of Chemistry, University of Tartu, Tartu, Estonia) / Alexandre Bastos (CICECO, DEMaC, University of Aveiro, Aveiro, Portugal) / Theodor Hack (Airbus Central R&T, Munich) / Mikhail Zheludkevich (Magnesium Innovation Centre–MagIC, Helmholtz-Zentrum Geesthacht (HZG), Geesthacht) / Mário Ferreira (CICECO, DEMaC, University of Aveiro, Aveiro, Portugal)	116267
WP1.9	Extraction of essential oil from Artemisia herba alba and it application in the protection of Aluminum in HCl medium	Nacer HECHICHE (université Mouloud Mammeri Tizi-Ouzou, Tizi-Ouzou, Algeria)	117932

Posternumber	Paper title	Authors	Submission ID
WP1.10	Surface treatments for controlling corrosion rate of biodegradable Mg alloy in simulated body fluid	Husnu Gerengi (Corrosion Research Laboratory, Department of Mechanical Engineering, Faculty of Engineering, Duzce University, 81620, Turkey) / Ertugrul Kaya (Corrosion Research Laboratory, Department of Mechanical Engineering, Faculty of Engineering, Duzce University, Duzce, Turkey)	118707
WP1.11	Analysis of Egyptian plant waste streams as a potential source of green corrosion inhibitors	Nicola Everitt (The University of Nottingham, Nottingham, United Kingdom) / Katy Voisey (The University of Nottingham, Nottingham, United Kingdom) / Irene Samy (Nile University, Cairo, Egypt)	119242
WP1.12	Effect of FeAl-alloy microstructure on the oxidation process of the intermetallic phase at 900 and 1000°C	Dorota Pasek (Silesian University of Technology, Katowice, Poland) / Maria Sozańska (Silesian University of Technology, Katowice, Poland) / Janusz Cebulski (Silesian University of Technology, Katowice, Poland)	119307
WP1.13	The influence of phase structure on corrosion behavior of magnesium based alloys with lithium addition	Ewa Ura-Binczyk (Warsaw University of Technology, Warsaw, Poland) / Anna Dobkowska (Warsaw Univerisity of Technology, Warsaw , Poland) / Boguslawa Adamczyk-Cieslak (Warsaw University of Technology, Warsaw, Poland) / Jaroslaw Mizera (Warsaw University, Warsaw, Poland)	119402
WP1.14	Continuous Downhole Corrosion Inhibitor Treatment of Selected Saudi Aramco Water Supply Wells	Yahya T. AlJanabi (Saudi Aramco, Dhahran , Saudi Arabia) / Mostafa M. Alsalem (Saudi Aramco, Dhahran, Saudi Arabia) / Mustafa S. Al-Jawad (Saudi Aramco, Dhahran, Saudi Arabia) / David R. Lewis (Saudi Aramco, Dhahran, Saudi Arabia)	119440
WP1.15	Dandelion extract as a green corrosion inhibitor for carbon steel in 0.5 M H2SO4 solution	Mohamed Deyab (Egyptian Petroleum Research Institute, cairo, Egypt)	120037
WP1.16	Effects of TiO2 nanoparticles on anti-corrosion performance of brass in simulated water containing surfactant	Hong-Hua Ge (Shanghai University of Electric Power, Shanghai , China PR) / Kai Wu (Shanghai University of Electric Power,, Shanghai, China PR)	120307
WP1.17	Water soluble Imidazolium Zwitterion as green corrosion inhibitors for mild steel: Experimental, DFT and MD studies	M. A. QURAISHI (Center of Research Excellence in Corrosion, Research Institute, King Fahd University of Petroleum and Minerals, Dhahran 31261, Saudi Arabia., Dhahran, Saudi Arabia) / Vandana Srivastava (Department of Chemistry, Indian Institute of Technology (Banaras Hindu University), Varanasi -221005, India., Varanasi, India) / Jiyaul Haque (Department of Chemistry, Indian Institute of Technology (Banaras Hindu University), Varanasi -221005, India., varanasi, India)	120357

Posternumber	Paper title	Authors	Submission ID
WP1.18	New carboxylate based corrosion inhibitors for mild steel.	Anthony Somers (Institute for Frontier Materials, Deakin University, Burwood, Australia) / Maria Forsyth (Institute for Frontier Materials, Deakin University, Burwood, Australia) / Esther Udabe (POLYMAT, Donostia-San Sebastian, Spain) / David Mecerreyes (POLYMAT, Donostia-San Sebastian, Spain)	121362

Corrosion by Hot Gases and Combustion Products (WP3)

Posternumber	Paper title	Authors	Submission ID
WP3.2	Increasing acid dew point corrosion resistance by thin coating layer on steel	Changhoon Choi (POSCO, 6262 Donghaean-ro, Nam-gu, Pohang-si, Korea, South) / Byoung Ho Lee (POSCO, Pohang-si, Korea, South) / Minho Jo (POSCO, Pohang-si, Korea, South) / Myung Soo Kim (POSCO, Gwangyang-si, Korea, South)	104977
WP3.3	Behavior of three cast chromium-rich cobalt-based alloys in oxidation at 900°C in a 100% water vapour atmosphere	Patrice Berthod (University of Lorraine / Institut Jean lamour, Nancy, France) / Mélissa Léglise (University of Lorraine / Institut Jean Lamour, Nancy, France) / Thierry Schweitzer (University of Lorraine / Institut Jean Lamour, Nancy, France) / Lionel Aranda (University of Lorraine / Institut Jean Lamour, Nancy, France) / Kevin Ginestar (Commissariat à l'Energie Atomique et aux Energies Alternatives, Saclay, France) / Sophie Bosonnet (Commissariat à l'Energie Atomique et aux Energies Alternatives, Saclay, France)	106567
WP3.4	Degradation at 550°C of cast chromium-rich cobalt-based alloys exposed to a H2O-O2-CO2-HCl gaseous mixture	Patrice Berthod (University of Lorraine / Institut Jean Lamour, Nancy, France) / lionel Aranda (University of Lorraine / Institut Jean Lamour, Nancy, France) / Thierry Schweitzer (University of Lorraine / Institut Jean Lamour, Nancy, France) / Christophe Rapin (University of Lorraine / Institut Jean Lamour, Nancy, France) / Didier Souchon (Veolia Environnement Recherche et Innovation, Centre de Recherche de Limay, Limay, France) / Farès Maad (Veolia Environnement Recherche et Innovation, Centre de Recherche de Limay, Limay, France) / Jean-Michel Brossard (Veolia Environnement Recherche et Innovation, Centre de Recherche de Limay, Limay, France)	106617
WP3.5	MATERIAL CORROSION TROUBLES ON HEAT RECOVERY SECTION IN WASTE TO ENERGY FACILITY	Ha-Na Jang (Yonsei University, Wonju, Korea, South) / Eun-song Lee (Yonsei University, Wonju, Korea, South)	107452

Posternumber	Paper title	Authors	Submission ID
WP3.6	The formation of ceramic coatings on the surface of plasma-chemical reactor for biomass utilization	Zydrunas Kavaliauskas (Lithuanian Energy Institute, Kaunas, Lithuania) / Dovilė Gimzauskaite (Lithuanian Energy Institute, Kaunas, Lithuania) / Viktorija Grigaitiene (Lithuanian Energy Institute, Kaunas, Lithuania) / Liutauras Marcinauskas (Lithuanian Energy Institute, Kaunas, Lithuania) / Vitas Valincius (Lithuanian Energy Institute, Kaunas, Lithuania) / MIndaugas Milieska (Lithuanian Energy Institute, Kaunas, Lithuania) / Romualdas Kezelis (Lithuanian Energy Institute, Kaunas, Lithuania)	107842
WP3.7	Hot corrosion of HP/HVOF sprayed MCrAlY coatings in molten salt environment	Kate¿ina Kopelentová (Výzkumný a zkušební ústav Plzeň s.r.o., Plzeň Plzeň, Czech Republic) / Zdeněk Česánek (Výzkumný a zkušební ústav Plzeň s.r.o., Pilsen, Czech Republic) / Jan Schubert (Výzkumný a zkušební ústav Plzeň s.r.o., Pilsen, Czech Republic)	109367
WP3.8	High temperature hot corrosion behavior of selected thermally sprayed coatings in an aggressive environment at 690 °C	Zdeněk Česánek (Výzkumný a zkušební ústav Plzeň s.r.o., Pilsen, Czech Republic) / Kateřina Kopelentová (Výzkumný a zkušební ústav Plzeň s.r.o., Pilsen, Czech Republic) / Jan Schubert (Výzkumný a zkušební ústav Plzeň s.r.o., Pilsen, Czech Republic)	109387
WP3.9	Gas phase conductivity and its effect to Cr2O3 volatility in Ar and Ar-5%H2	Syamsul Kamal Arifin (International Islamic University Malaysia, Kuala Lumpur, Malaysia) / Ahmad Nukhaie Berahim (International Islamic University Malaysia, Kuala Lumpur, Malaysia) / Mohd Hanafi Ani (International Islamic University Malaysia, Kuala Lumpur, Malaysia)	110297
WP3.10	Current-Voltage measurement of Fe-Cr alloy at high temperature in dry and wet condition	Ahmad Nukhaie Berahim (International Islamic University Malaysia, Kuala Lumpur, Malaysia) / Syamsul Kamal Arifin (International Islamic University Malaysia, Kuala Lumpur, Malaysia) / Muhammad Syamim Mohamad Fauzi (International Islamic University Malaysia, Kuala Lumpur, Malaysia) / Mohd Hanafi Ani (International Islamic University Malaysia, Kuala Lumpur, Malaysia)	110922

Posternumber	Paper title	Authors	Submission ID
WP3.11	Microstructure of Sanicro 25 after long-term steam oxidation characterised by advanced electron microscopy techniques	Grzegorz Cempura (1AGH University of Science and Technology, International Centre of Electron Microscopy for Materials Science and Faculty of Metals Engineering and Industrial Computer Science, Krakow, Poland) / Aleksander Gil (2AGH University of Science and Technology, Faculty of Materials Science and Ceramics, Krakow, Poland) / Alina Aguero (Instituto Nacional de Técnica Aeroespacial, Departamento de Materiales y Estructuras, Torrejón de Ardoz, Spain) / Adam Kruk (1AGH University of Science and Technology, International Centre of Electron Microscopy for Materials Science and Faculty of Metals Engineering and Industrial Computer Science, Kraków, Poland) / Aleksandra Czyrska-Filemonowicz (1AGH University of Science and Technology, International Centre of Electron Microscopy for Materials Science and Faculty of Metals Engineering and Industrial Computer Science, Kraków, Poland) / Marcos Gutierrez (Instituto Nacional de Técnica Aeroespacial (INTA), Departamento de Materiales y Estructuras, Torrejón de Ardoz, Spain)	113982
WP3.12	The effect of reactive elements on the formation and hot corrosion performance of aluminide coating	Farhad Shahriari Nogorani (Shiraz University of Technology, Shiraz, Iran) / Farhad Fatemi (Shiraz University of Technology , Shiraz, Iran)	115107
WP3.13	Anti-corrosion coatings produced from complex oxides	Oleg Chizhko (German-Russian College, Tscherkessk, Russia)	118347
WP3.14	Lifetime Prediction of Yttria Stabilized Zirconia (YSZ), La2Zr2O7, Gd2Zr2O7, YSZ/ La2Zr2O7 and YSZ/ Gd2Zr2O7 Thermal Barrier Coatings (TBCs) with Cold Gas Dynamic Spray (CGDS) bond coat under Furnace Thermal Cyclic Loadings	Abdullah Cahit Karaoglanli (Bartin University, Department of Metallurgical and Materials Engineering, Bartin, Turkey) / Kadir Mert Doleker (Bartin University, Department of Metallurgical and Materials Engineering, Bartin, Turkey)	122035
WP3.15	Investigation of Microstructure, Oxidation, and Interface Properties of Single and Double-Layer Gadolinium Zirconate/Yttria Stabilized Zirconia Thermal Barrier Coatings (TBCs)	Abdullah Cahit Karaoglanli (Bartin University, Department of Metallurgical and Materials Engineering, Bartin, Turkey) / Kadir Mert Doleker (Bartin University, Department of Metallurgical and Materials Engineering, Bartin, Turkey)	122050
Nuclear Corrosion (WP4)			
WP4.3	Investigation of the effects of irradiation on corrosive environments using in-situ electrochemical measurements during gamma-ray irradiation	Tomonori Sato (Japan Atomic Energy Agency, Tokai-mura, Japan) / Fumiyoshi Ueno (Japan Atomic Energy Agency, Tokai-mura, Japan)	115387
WP4.5	Corrosion Characteristics of Additively Manufactured Engineering Metallic Components	Raúl B. Rebak (GE Global Research, Schenectady, United States of America) / Xiaoyuan Lou (Auburn University, Auburn, United States of America)	121217

Posternumber	Paper title	Authors	Submission ID
WP4.6	High temperature corrosion phenomena during a nuclear severe accident – The corium-concrete interaction	Andrea Quaini (Den-Service de la Corrosion et du Comportement des Matériaux dans leur Environnement (SCCME), CEA, Université Paris-Saclay, Gif-sur-Yvette, France) / Christine Guéneau (Den-Service de la Corrosion et du Comportement des Matériaux dans leur Environnement (SCCME), CEA, Université Paris-Saclay , Gif-sur-Yvette, France) / Stephane Gossé (Den-Service de la Corrosion et du Comportement des Matériaux dans leur Environnement (SCCME), CEA, Université Paris-Saclay , Gif-sur-Yvette, France) / Fiqiri Hodaj (Univ. Grenoble Alpes, CNRS, Grenoble INP, SIMAP, Grenoble, France)	121610

Environment Sensitive Fracture (WP5)

Posternumber	Paper title	Authors	Submission ID
WP5.1	Stress corrosion cracking behavior of E690 steel in thiosulfate-containing artificial seawater	Huiyun Tian (Ocean University of China, Qingdao, China PR) / Qiankun Lu (Ocean University of China, Qingdao, China PR) / Zhongyu Cui (Ocean University of China, Qingdao, China PR) / Xin Wang (Ocean University of China, Qingdao, China PR)	104657
WP5.2	Diffusivity of hydrogen in advanced high strength steels and model structures	Václav Šefl (University of Chemistry and Technology Prague, Technopark Kralupy, Kralupy nad Vltavou, Czech Republic) / Darya Rudomilova (University of Chemistry and Technology Prague, Technopark Kralupy, Kralupy nad Vltavou, Czech Republic) / Tomáš Prošek (University of Chemistry and Technology Prague, Technopark Kralupy, Kralupy nad Vltavou, Czech Republic) / Pavel Salvetr (University of Chemistry and Technology Prague, Prague, Czech Republic) / Pavel Novák (University of Chemistry and Technology Prague, Prague, Czech Republic) / Andreas Muhr (voestalpine Stahl GmbH, Linz, Austria) / Gerald Luckeneder (voestalpine Stahl GmbH, Linz, Austria) / Hubert Duchaczek (voestalpine Stahl GmbH, Linz, Austria)	109322
WP5.4	HYDROGEN UPTAKE AND ITS EFFECTS ON MECHANICAL PERFORMANCE OF NODULAR CAST IRON	Antti Forsström (Aalto University School of Engineering,, Espoo, Finland) / Hannu Hanninen (Aalto University School of Engineering,, Espoo, Finland) / Yuriy Yagodzinskyy (Aalto University School of Engineering, Espoo, Finland)	118687
WP5.5	Environmental Testing of Wrought and Hot Isostatically Pressed Duplex Stainless Steels	Lisa Blanchard (University of Leicester, Leicester, United Kingdom) / Kasra Sotoudeh (TWI Ltd, Cambridge, United Kingdom) / Hongbiao Dong (University of Leicester, Leicester, United Kingdom)	121412

Posternumber	Paper title	Authors	Submission ID
	Investigating Hydrogen Embrittlement in a Nano-structured Steel	Arash Shadravan (Texas A&M University , College Station, United States of America) / Raymundo Case (Texas A&M University, College Station, United States of America)	117107

Corrosion mechanisms, methods and modelling (WP6 & WP8)

Posternumber	Paper title	Authors	Submission ID
WP6-8.1	The application of ENA for evaluation of the protective performance of ZRP modified by combinations of Zn dust with stainless steel flakes type of filler	Boleslav Eremias (SVUOM Ltd., Prague, Czech Republic) / Lubomir Mindos (SVUOM Ltd., Prague, Czech Republic) / Libor Turek (SVUOM Ltd., Prague, Czech Republic) / Libuse Hochmannova (SYNPO Inc., Pardubice, Czech Republic)	101467
WP6-8.2	ASTM-A-763 vs Ecorr. Test variables influence in ferritic stainless steel.	Victoria Matres (Acerinox Europa S.A.U., Palmones-Los Barrios, Spain) / Ana Notario (Acerinox Europa, Palmones-Los Barrios, Spain)	101577
WP6-8.3	Investigation of corrosion for aluminum fin-tube heat exchanger according to the corrosion potential and solution conductivity using a boundary element method	Yong-Sang Kim (Sungkyunkwan University, Suwon, Korea, South) / Jung-Gu Kim (Sungkyunkwan University, Suwon, Korea, South)	101771
WP6-8.4	Effect of water content on corrosion of mild steel in DESs in conditions of natural aeration	Yelyzaveta Rublova (Ukrainian State University of Chemical Technology, Dnipro, Ukraine) / Anna Kityk (Ukrainian State University of Chemical Technology, Dnipro, Ukraine) / Natalia Bannyk (Ukrainian State University of Chemical Technology, Dnipro, Ukraine) / Valerii Malyshev (Institute of Physical Chemistry PAS, Warsaw, Poland) / Anna Kelm (Institute of Physical Chemistry PAS, Warsaw, Poland)	104132
WP6-8.5	Dependence of pitting of duplex stainless steels on potential and temperature	Yongsun Yi (Khalifa University, Abu Dhabi, United Arab Emirates) / Pyungyeon Cho (Khalifa University, Abu Dhabi, United Arab Emirates) / Sara Al Saadi (Federal Authority for Nuclear Regulation, Abu Dhabi, United Arab Emirates) / Changheui Jang (Korea Advanced Institute of Science & Technology , Daejeon, Korea, South)	104407
WP6-8.6	Study of corrosion inhibitors on AZ91D magnesium alloy	Yan li (Northeastern University , Shenyang, China PR) / Xiaopeng Lu (Northeastern University, Liaoning, China PR) / Tao Zhang (Northeastern University, Shenyang, China PR) / Fuhui Wang (Northeastern University, Shenyang, China PR)	105322

Posternumber	Paper title	Authors	Submission ID
WP6-8.7	On the mechanism of the corrosion of Al2Cu intermetallic phase in the acidic solution	Przemyslaw Kwolek (Rzeszow University of Technology, Faculty of Mechanical Engineering and Aeronautics, Department of Materials Science, Rzeszow, Poland) / Andrzej Gradzik (Rzeszow University of Technology, Faculty of Mechanical Engineering and Aeronautics, Department of Materials Science, Rzeszow, Poland) / Dariusz Szeliga (Rzeszow University of Technology, Faculty of Mechanical Engineering and Aeronautics, Department of Materials Science, Rzeszow, Poland)	106687
WP6-8.8	Influence of the driving force of diffusion and kinetic demixing processes on the oxidation kinetics of Ni,Cr alloys and CaO or CeO2 coated nickel.	Nacer Halem (University, TIZI OUZOU, Algeria) / ZOHRA HALEM (University, BOUIRA, Algeria) / LUKASZ CIENIEK (University, KRAKOW, Poland) / JEAN KUSINSKI (University, KRAKOW, Poland) / GEORGETTE PETOT-ERVAS (University, PARIS, France)	107292
WP6-8.9	Effects of Na2MoO4 in chemical passivation surface treatment on the passive films of low nickel duplex stainless steel	Jongbeom Choi (Korea Maritime and Ocean University, Busan, Korea, South) / Jeonghyeon Yang (Gyeongsang National University, Tongyeong, Korea, South) / Jun Kang (Korea Maritime and Ocean University, Busan, Korea, South) / Myeonghoon Lee (Korea Maritime and Ocean University, Busan, Korea, South) / Yongsup Yun (Korea Maritime and Ocean University, Busan, Korea, South) / kyunghwang lee (Research Institute of Industrial Science & Technology, Materials Solution Research Group, pohang, Korea, South)	107557
WP6-8.10	CFD simulation of CO2 corrosion under flow conditions in an elbow	Jukai Chen (State Key Laboratory of Multiphase Flow in Power Engineering, Xi'an Jiaotong University, Xi'an, China PR) / Xiaodan Wang (State Key Laboratory of Multiphase Flow in Power Engineering, Xi'an Jiaotong University, Xi'an, China PR) / Yueshe Wang (State Key Laboratory of Multiphase Flow in Power Engineering, Xi'an Jiaotong University, Xi'an, China PR)	107742
WP6-8.11	Inverse Methods in Corrosion of Rebars in Concrete	Robert Filipek (AGH-University of Science and Technology, Faculty of Materials Science and Ceramics, Krakow, Poland) / Krzysztof Szyszkiewicz-Warzecha (AGH-University of Science and Technology, Faculty of Materials Science and Ceramics, Kraków, Poland)	108872

Posternumber	Paper title	Authors	Submission ID
WP6-8.12	Exploring lithium conversion layer formation on AA2024-T3 by TEM-EELS/EDS	Ali Kosari (Department of Materials Science and Engineering, Delft University of Technology, Delft, Netherlands) / Peter Visser (Research Development and Innovation, AkzoNobel, Sassenheim, Netherlands) / Peyman Taheri (Department of Materials Science and Engineering, Delft University of Technology, Delft, Netherlands) / Henny Zandbergen (Kavli Institute of Nanoscience, Delft University of Technology, Delft, Netherlands) / Frans Tichelaar (Kavli Institute of Nanoscience, Delft University of Technology, Delft, Netherlands) / Herman Terryn (Department of Materials and Chemistry, Research Group Electrochemical and Surface Engineering (SURF), Vrije Universiteit Brussel, Brussels, Netherlands) / Arjan Mol (Department of Materials Science and Engineering, Delft University of Technology, Delft, Netherlands)	109512
WP6-8.13	Study of cupric ion effect in Nickel-Chromium multilayer systems corrosion by localised techniques	Eva Garcia-Lecina (CIDETEC , San Sebastián, Spain) / Larraitz Ganborena (CIDETEC , San Sebastián, Spain) / Jesús Manuel Vega (CIDETEC, San Sebastián, Spain) / Hans-Jürgen Grande (CIDETEC , San Sebastián, Spain)	109712
WP6-8.14	Corrosion study in artificial saliva of dental alloys based on cobalt and chromium, using electrochemical stationary methods and EIS	Patrice Berthod (University of Lorraine / Faculty of Sciences and Technologies / Institut Jean Lamour, Vandoeuvre-lès Nancy, France) / Xin Li (University of Lorraine / Faculty of Sciences and Technologies. Vandoeuvre-lès-Nancy. France) / Estelle Kretz	109942
	Agar-based gels as an electrolyte for corrosion diagnostics	Andreas Heyn (Otto-von-Guericke-Universität Magdeburg, Magdeburg)	109987
WP6-8.16	Proposal of a model based on Tafel-Piontelli laws for the calculation of corrosion rate of metals in acidic condition	Andrea Brenna (Politecnico di Milano, Dipartimento di Chimica, Materiali e Ingegneria Chimica "Giulio Natta", Milano, Italy) / Fabio Bolzoni (Politecnico di Milano, Dipartimento di Chimica, Materiali e Ingegneria Chimica "Giulio Natta", Milano, Italy) / Cristina De Giovanni (Politecnico di Milano, Dipartimento di Chimica, Materiali e Ingegneria Chimica "Giulio Natta", Milano, Italy) / Marco Ormellese (Politecnico di Milano, Dipartimento di Chimica, Materiali e Ingegneria Chimica "Giulio Natta", Milano, Italy)	111282
WP6-8.17	Age-hardening behaviour, microstructure and corrosion resistance of the copper alloyed stainless steel 1.4542	Norman Kauss (Otto-von-Guericke Universität Magdeburg, Magdeburg) / Paul Rosemann (Bundesanstalt für Materialforschung und –prüfung (BAM), Berlin) / Thorsten Halle (Otto-von-Guericke Universität, Magdeburg)	111777

Posternumber	Paper title	Authors	Submission ID
WP6-8.18	Effect of cryogenic treatment on Corrosion behaviour of ultrafine-grained AA6012-T6 alloy processed by ECAP	Annamaria Viceré (Università Politecnica delle Marche, Ancona, Italy) / Tiziano Bellezze (Università Politecnica delle Marche, Ancona, Italy) / Chiara Paoletti (Università Politecnica delle Marche, Ancona, Italy) / Marcello Cabibbo (Università Politecnica delle Marche, Ancona, Italy) / Gabriella Roventi (Università Politecnica delle Marche, Ancona, Italy)	111837
WP6-8.19	Visualization of material-related susceptibility to pitting corrosion using the "KorroPad" indicator test	Paul Rosemann (Federal Institute for Materials Research and Testing, Berlin) / Norman Kauss (Institut für Werkstoff- und Fügetechnik, Otto-von-Guericke University, Magdeburg)	113312
WP6-8.20	Importance of keeping surfaces clean: root causes of early corrosion of copper tubing	Elena Mielgo (ITMA Materials Technology, Avilés, Spain) / Olga Conejero (ITMA Materials Technology, Avilés, Spain)	113937
WP6-8.21	Estimate metals leaching from stainless steel cookware into food using chemical and electrochemical processes	Abdel Hamid Zeinab (CMRDI, He, Egypt)	114137
WP6-8.22	KorroPad - A new test method for the investigation on the passive layer stablity on stainless steels	Thoralf Müller (Bundesanstalt für Materialforschung und -prüfung (BAM), Berlin) / Jens Lehmann (Bundesanstalt für Materialforschung und -prüfung (BAM), Berlin) / Andreas Burkert (Bundesanstalt für Materialforschung und -prüfung (BAM), Berlin)	114917
WP6-8.23	Modelling the Electrochemical Interface by DFT – the Case of Aqueous Copper Corrosion	Joakim Halldin Stenlid (Fysikum, Stockholm University, Stockholm, Sweden) / A Johannes Johansson (Swedish Nuclear Fuel and Waste Management Co (SKB), Solna, Sweden) / Lars GM Pettersson (Fysikum, Stockholm University, Stockholm, Sweden)	115972
WP6-8.24	Precipitation behavior and corrosion resistance of nickel-free, high-nitrogen austenitic stainless steels	Paul Rosemann (Federal Institute for Materials Research and Testing, Berlin)	116137
WP6-8.25	On the effect of detwinning on corrosion behavior of NiTi shape memory alloy	Mahdi Mohajeri (Department of Materials Science and Engineering, Texas A&M University, College Station, United States of America) / Behrouz Haghgouyan (Department of Materials Science and Engineering, Texas A&M University, College Station, United States of America) / Homero Castaneda (Department of Materials Science and Engineering, Texas A&M University, College Station, United States of America) / Dimitris C. Lagoudas (Department of Aerospace Engineering, Texas A&M University, College Station, United States of America)	116567

Posternumber	Paper title	Authors	Submission ID
WP6-8.26	Comparison of the corrosion resistance of AA2024 and AA2098 alloys in different solutions	Mariana Xavier Milagre (Instituto de Pesquisas Energéticas e Nucleares, São Paulo, Brazil) / João Victor Araujo (Instituto de Pesquisas Energéticas e Nucleares, São Paulo, Brazil) / Caruline de Souza Carvalho Machado (Intituto de Pesquisas Energéticas e Nucleares, São Paulo, Brazil) / Uyime Donatus (Intituto de Pesquisas Energéticas e Nucleares, São Paulo, Brazil) / Vishnu Mogili (Laboratorio Nacional de Nanotecnologia, Campinas, Brazil) / isolda Costa (Instituto de Pesquisas Energéticas e Nucleares, São Paulo, Brazil) / Maurilio Pereira Gomes (Instituto de Pesquisas Energéticas e Nucleares, São Paulo, Brazil)	116592
WP6-8.27	An experimental investigation on corrosion behavior of NiTi shape memory alloy under uniaxial tension	Mahdi Mohajer (Department of Materials Science and Engineering, Texas A&M University, College Station, United States of America) / Behrouz Haghgouyan (Department of Materials Science and Engineering, Texas A&M University, College Station, United States of America) / Homero Castaneda (Department of Materials Science and Engineering, Texas A&M University, College Station, United States of America) / Dimitris C. Lagoudas (Department of Aerospace Engineering, Texas A&M University, College Station, United States of America)	116627
WP6-8.28	Corrosion behavior of friction stir welded 2198-T8 aluminum-copper-lithium alloy	Caruline de Souza Carvalho Machado (IPEN, São Paulo, Brazil) / Mariana Xavier Milagre (IPEN, São Paulo, Brazil) / Rafael Emil Klumpp (IPEN, São Paulo, Brazil) / João Victor Araujo (IPEN, São Paulo, Brazil) / Vishnu Mogili (LnNANO, Campinas, Brazil) / Rafael A.R. GIORJAO (LnNANO, Campinas, Brazil) / Isolda Costa (IPEN, São Paulo, Brazil) / Uyme Donatus (IPEN, São Paulo, Brazil) / Victor Ferrinho Pereira (LNNano/CNPEM, Campinas, Brazil)	116767
WP6-8.29	Application of Cyclic Thermal-oxidized IrOx electrode in pH detection of interface region of Zn/Steel galvanic process	Feifei Huang (University of Science and Technology Beijing, Beijing, China PR) / Ying Jin (University of Science and Technology Beijing, Beijing, China PR) / Lei Wen (University of Science and Technology Beijing, Beijing, China PR)	117472
WP6-8.30	The partial processes of anodic oxidation of zinc in sodium hydroxide	Irina V. Protasova (Voronezh State University, Voronezh, Russia)	117987
WP6-8.31	CO2 solubility measurement and its effect on the corrosion	Yuanhua Lin (Southwest Petroleum University, Chengdu, China PR)	118062

Posternumber	Paper title	Authors	Submission ID
WP6-8.32	Electrodeposition and the corrosion resistance of nanocrystalline zinc coating on steel substrates	Maozhong An (Harbin Institute of Technology, Haerbin, China PR) / Xiaochuan Ma (Harbin Institute of Technology, Harbin, China PR) / Qingyang Li (Harbin Institute of Technology, Harbin, China PR) / Jinqiu Zhang (Harbin Institute of Technology, Harbin, China PR) / Peixia Yang (Harbin Institute of Technology, Harbin, China PR)	118602
WP6-8.33	Effect of additives on the corrosion mechanism of nanocrystalline zinc–nickel alloys in an alkaline bath	PEIXIA YANG (Harbin Institute of Technology, Harbin , China PR) / DAN WANG (Harbin Institute of Technology, Harbin, China PR) / ZHONGBAO FENG (Harbin Institute of Technology, Harbin, China PR) / JINQIU ZHANG (Harbin Institute of Technology, Harbin, China PR) / MAOZHONG AN (Harbin Institute of Technology, Harbin, China PR)	118632
WP6-8.34	Corrosion of cast magnesium alloys in sulphuric acid used in removal of casting skin contamination	Adam Gryc (Silesian University of Technology, Katowice, Poland) / Andrzej Kiełbus (Silesian University of Technology, Katowice, Poland) / Maria Sozańska (Silesian University of Technology, Katowice, Poland)	118807
WP6-8.35	STUDIES ON THE CORROSION RESISTANCE OF LASER-WELDED 304 AND 304H STEEL	Katarzyna Lyczkowska (Politechnika Slaska, Katowice, Poland) / Janusz Adamiec (Politechnika Slaska, Katowice, Poland) / Maria Sozanska (Politechnika Slaska, Katowice, Poland) / Joanna Michalska (Politechnika Slaska, Gliwice, Poland)	118867
WP6-8.36	Test Of A High Resolution Electrical Resistance Corrosion Sensor For Industrial Aplications	Otávio Carneiro Corrêa (Vidya Technology, Curitiba, Brazil) / Fulvio Silva (Vidya Technology, Curitiba, Brazil)	119960
WP6-8.37	Corrosion behaviour in sodium chloride solution of sites containing sigma phase in super duplex stainless steels	Paulina Erazmus Vignal (SATT Grand Est, Dijon, France) / vincent vignal (ICB, Dijon, France) / Saghi Saedlou (APERAM R&D, isbergues, France) / Florent Krajcarz (APERAM R&D, Isbergues, France)	120242
WP6-8.38	Artificial neural networks combined with sensitivity analysis as a prediction tool for breakdown potential values of stainless steel in pitting corrosion modelling	María Jesús Jiménez come (Universidad de Cádiz, Algeciras Algeciras, Spain) / Maria de la Luz Martín (Universidad de Cádiz, Algeciras, Spain) / Francisco J. Trujillo Espinosa (Universidad de Cádiz, Algeciras, Spain) / Victoria Matres Serrano (ACERINOX EUROPA S.A.U, Los Barrios, Spain)	120422
WP6-8.39	The New Potential in Thermal Diffusion Galvanizing	Anna Wassilkowska (Cracow University of Technology, Kraków, Poland) / Alexander Biryukov (Chelyabinsk State University, Chelyabinsk , Poland) / Dmitry Zakharyevich (Chelyabinsk State University , Chelyabinsk , Poland) / Rashid Galin (Vika-Gal2 Ltd. , Chelyabinsk , Poland)	120695

Posternumber	Paper title	Authors	Submission ID
WP6-8.40	Visualizing Corrosion Mechanisms on Galvanized Steel	Markus Nadlinger (CEST, 2700 Wiener Neustadt, Austria) / Martin Fleischanderl (voestalpine Stahl GmbH, 4030 Linz, Austria) / Martin Rosner (voestalpine Stahl GmbH, 4030 Linz, Austria) / Christian Paulik (JKU, Institute for Chemical Technology of Organic Materials , 4040 Linz, Austria) / Karl-Heinz Stellnberger (voestalpine Stahl GmbH, 4030 Linz, Austria) / Gerald Luckender (voestalpine Stahl GmbH, 4030 Linz, Austria)	121042
WP6-8.41	Selective Oxidation of Fe-Mn (1 wt.%) Binary Alloy	Li Gong (CentraleSupélec-- Universite Paris Saclay, 91190 GIF SUR YVETTE, France) / Nathalie Ruscassier (CentraleSupélec-- Universite Paris Saclay, 91190 GIF SUR YVETTE, France) / Mehdi Ayouz (CentraleSupélec-- Universite Paris Saclay, 91190 GIF SUR YVETTE, France) / Marie-Laurence Giorgi (CentraleSupélec-- Universite Paris Saclay, 91190 GIF SUR YVETTE, France)	121287
WP6-8.42	Investigation the biocompatibility of nickel free lean Duplex Stainless Steel in Simulated body fluid	Ihsan Ul Haq Toor (KFUPM, Dhahran, Saudi Arabia) / farid Abdallah (KFUPM, Dhahran, Saudi Arabia)	121347
WP6-8.43	Atom probe tomography as a tool to analyze outward diffusion through Al2O3 on FeAlCr	Oscar Recalde (Institute for Applied Materials (IAM-WK), Karlsruhe Institute of Technology (KIT), 76344 Eggenstein-Leopoldshafen) / Alexander Kauffmann (Institute for Applied Materials (IAM-WK), Karlsruhe Institute of Technology (KIT), 76344 Eggenstein-Leopoldshafen) / Martin Heilmaier (Institute for Applied Materials (IAM-WK), Karlsruhe Institute of Technology (KIT), 76344 Eggenstein-Leopoldshafen) / Torben Boll (Karlsruhe Nano Micro Facility (KNMF), Karlsruhe Institute of Technology (KIT), 76344 Eggenstein-Leopoldshafen)	121387
WP6-8.44	Optimisation of the corrosion rate of iron-based alloys for bioresorbable stent applications by surface acidification	Sarah Reuter (Université catholique de Louvain, Institute of Mechanics, Materials and Civil Engineering, IMAP, Louvain-la-Neuve, Belgium) / Cédric Georges (CRM group, Liège, Belgium) / Pascal J Jacques (Université catholique de Louvain, Institute of Mechanics, Materials and Civil Engineering, IMAP, Louvain-la-Neuve, Belgium)	121432
WP6-8.45	Diffusion and migration measurement in a minimized migration cell with electrochemical methods	Lars Varain (TU Wien, Institute of Chemical Technologies and Analytics, Vienna, Austria) / Silvia Larisegger (KAI Kompetenzzentrum Automobil- und Industrieelektronik GmbH, Villach, Austria) / Michael Nelhiebel (KAI Kompetenzzentrum Automobil- und Industrieelektronik GmbH, Villach, Austria) / Günter Fafilek (TU Wien, Institute of Chemical Technologies and Analytics, Vienna, Austria)	121630

Posternumber	Paper title	Authors	Submission ID
WP6-8.46	A Novel X-Ray Diffraction Technique for in-situ Observations of Cathode/Anode Reaction in Electrolytes	Sabine Reither (KAI, Center of Expertise in Automotive and Industrial Electronics, Villach, Austria) / Silvia Larisegger (KAI, Center of Expertise in Automotive and Industrial Electronics , Villach, Austria) / Michael Nelhiebel (KAI, Center of Expertise in Automotive and Industrial Electronics , Villach, Austria) / Günter Fafilek (Institute of Chemical Technologies and Analytics, TU Wien, Vienna, Austria)	121655
WP6-8.47	Finite element analysis of the effect of interfacial bubbles on performance of epoxy coatings under the alternating hydrostatic pressure	Rui Liu (Corrosion and Protection Division, Shenyang National Laboratory for Materials Science, Institute of Metal Research, Chinese Academy of Sciences, Shenyang, China PR) / Li Liu (Corrosion and Protection Division, Shenyang National Laboratory for Materials Science, Institute of Metal Research, Chinese Academy of Sciences, Shenyang, China PR) / Wenliang Tian (Corrosion and Protection Division, Shenyang National Laboratory for Materials Science, Institute of Metal Research, Chinese Academy of Sciences, Shenyang, China PR) / Yu Cui (Corrosion and Protection Division, Shenyang National Laboratory for Materials Science, Institute of Metal Research, Chinese Academy of Sciences, Shenyang, China PR) / Ying Li (Corrosion and Protection Division, Shenyang National Laboratory for Materials Science, Institute of Metal Research, Chinese Academy of Sciences, Shenyang, China PR) / Fuhui Wang (Corrosion and Protection Division, Shenyang National Laboratory for Materials Science, Institute of Metal Research, Chinese Academy of Sciences, Shenyang, China PR)	121865
	In-situ investigation of the semiconductive properties and protective role of Cu2O layer formed on copper in a borate buffer solution	Pan Yi (Corrosion and Protection Center, University of Science and Technology Beijing, Beijing, China PR) / Chaofang Dong (Corrosion and Protection Center, University of Science and Technology Beijing, Beijing, China PR) / Kui Xiao (Corrosion and Protection Center, University of Science and Technology Beijing, Beijing, China PR) / Cheng Man (Corrosion and Protection Center, University of Science and Technology Beijing, Beijing, China PR) / Xiaogang Li (Corrosion and Protection Center, University of Science and Technology Beijing , Beijing, China PR)	122300

Corrosion Education (WP7)

WP7.1	Heat treatment and corrosion resistance of cutlery	Paul Rosemann (Federal Institute for Materials Research and Testing, Berlin)	116037

Posternumber	Paper title	Authors	Submission ID
Marine Corrosion (WP9)			
WP9.1	An investigation into Intelligent Self-Monitoring Coating Systems	Craig Melton (The University of Leicester, Leicester, United Kingdom) / Simon Gill (The University of Leicester, Leicester, United Kingdom) / Shiladitya Paul (The Welding Institute, Cambridge, United Kingdom)	102622
WP9.2	Stress corrosion cracking behaviors of E690 steel in SO2-polluted marine atmosphere	Hongchi Ma (University of Science & Technology Beijing, Beijing, China PR) / Zhiyong Liu (University of Science & Technology Beijing, Beijing, China PR) / Cuiwei Du (University of Science & Technology Beijing, Beijing, China PR) / Xiaogang Li (University of Science & Technology Beijing, Beijing, China PR)	103182
WP9.3	Effect of hydrostatic pressure on metastable pitting corrosion of X70 pipeline steel in containing chlorine ion solution	Jinxu Li (University of Science and Technology Beijing, Beijing, China PR) / Zixuan Yang (University of Science and Technology Beijing, Beijing, China PR)	104837
WP9.4	Transparent Acrylic Coating Protection of Brass in a Marine Atmosphere	Olga Papadopoulou (National Technical University of Athens, Athens, Greece) / Stefanos Xifolis (National Technical University of Athens, Athens, Greece) / Panayota Vassiliou (National Technical University of Athens, Athens, Greece)	115762
Microbial Corrosion (WP10)			
WP10.1	Corrosion of Q235 carbon steel affected by the introduction of aerogenic and aerobic bacteria	Dun Zhang (Institute of Oceanology, Chinese Academy of Sciences, Qingdao, China PR)	101567
WP10.2	Laboratory investigation of microbiologically influenced corrosion of Q235 carbon steel by archaea Natronorubrum tibetense	Hongchang Qian (University of Science and Technology Beijing, Beijing, China PR) / Dawei Zhang (University of Science and Technology Beijing, Beijing, China PR) / Dake Xu (Northeastern University, Shenyang, China PR) / Cuiwei Du (University of Science and Technology Beijing, Beijing, China PR) / Xiaogang Li (University of Science and Technology Beijing, Beijing, China PR)	104377
WP10.3	Facile in situ growth of high strong BiOI network films on metal wire meshes with photocatalytic activity	Yi Wang (Institute of Oceanology, Chinese Academy of Sciences, Qingdao, China PR) / Dun Zhang (Institute of Oceanology, Chinese Academy of Sciences, Qingdao, China PR)	104732
WP10.4	The impact of microorganisms on the corrosion protection by self-assembled layers of phosphonic acids in natural waters	Ekatarina Kristan Mioč (Faculty of Chemical Engineering and Technology, University of Zagreb, Zagreb, Croatia) / Helena Otmačić Ćurković (Faculty of Chemical Engineering and Technology, University of Zagreb, Zagreb, Croatia)	106132

Posternumber	Paper title	Authors	Submission ID
WP10.5	SECM investigation of the effect of riboflavin secretion of Shewanella oneidensis on steel corrosion	Ziyu Li (University of Science and Technology Beijing, Beijing, China PR) / Dawei Zhang (University of Science and Technology Beijing, Beijing, China PR) / Dake Xu (Northeastern University, Shenyang, China PR) / Xiaogang Li (University of Science and Technology Beijing, Beijing, China PR)	107512
WP10.6	The pH Effect on Microbiological Influence Corrosion in Circulated Water Systems	Wei-Jen Li (Industrial Technology Research Institute, Hsinchu, Taiwan)	107732
WP10.7	Functionalised new gemini surfactants as biocorrosion and corrosion inhibitors	Marta Pakiet (Adam Mickiewicz University in Poznan, Faculty of Chemistry, Poznań Poznań, Poland) / Iwona Kowalczyk (Adam Mickiewicz University in Poznan, Poznań, Poland) / Rafael Leiva Garcia (University of Manchester, Manchester, United Kingdom) / Robert Moorcroft (University of Manchester, Manchester, United Kingdom) / Tim Nichol (Sheffield Hallam University, Sheffield, United Kingdom) / Tom Smith (Sheffield Hallam University, Sheffield, United Kingdom) / Robert Akid (University of Manchester, Manchester, United Kingdom) / Bogumił Brycki (Adam Mickiewicz University in Poznań, Poznań, Poland)	108447
WP10.8	Corrosion behavior of different copper alloys in the food industry in presence of Pseudomonas fluorescens	Giorgia Ghiara (Università di Genova, genova, Italy) / Roberto Spotorno (Università di Genova, Genova, Italy) / Pierangela Cristiani (RSE SpA - CNR, Institute of Biomolecular Chemistry, Milano, Italy)	109552
WP10.9	Corrosion of orthodontic archwires in artificial saliva in the presence of Lactobacillus reuteri	Helena Otmacic-Curkovic (Faculty of Chemical Engineering and Technology, University of Zagreb, Zagreb, Croatia) / Nikolina Leona Serdarević (Faculty of Chemical Engineering and Technology, University of Zagreb, Zagreb, Croatia) / Ines Musa Trolic (School of Dental Medicine, University of Zagreb, Zagreb, Croatia) / Stjepan Spalj (School of Medicine, University of Rijeka, Rijeka, Croatia) / Zrinka Todoric (School of Medicine, University of Zagreb, Zagreb, Croatia) / Ana Budimir (School of Medicine, University of Zagreb, Zagreb, Croatia)	112597

Posternumber	Paper title	Authors	Submission ID
WP10.10	Assessment of Ag-doped hybrid coatings for microbial corrosion protection of steel	Leonardo Iannucci (Dipartimento di Elettronica e Telecomunicazioni, Politecnico di Torino, Torino, Italy) / Emma Angelini (Dipartimento di Scienza Applicata e Tecnologia, Politecnico di Torino, Torino, Italy) / Sabrina Grassini (Dipartimento di Scienza Applicata e Tecnologia, Politecnico di Torino, Torino, Italy) / Marco Parvis (Dipartimento di Elettronica e Telecomunicazioni, Politecnico di Torino, Torino, Italy) / Marco Sangermano (Dipartimento di Scienza Applicata e Tecnologia, Politecnico di Torino, Torino, Italy) / Pierangela Cristiani (Ricerca sul Sistema Energetico (RSE SpA), Milano, Italy)	114842
WP10.11	Corrosion of carbon steel in the synthetic bentonite pore water inoculated with a natural microbial consortium dominated by SRB	Jana Steinová (Technical University Liberec, Liberec, Czech Republic) / Tomáš Černoušek (Research CentreŘež, Husinec-Řež, Czech Republic) / Hana Kovářová (Research Centre Řež, Husinec-Řež, Czech Republic) / Roman Špánek (Technical University Liberec, Liberec, Czech Republic) / Jakub Kokinda (Research Centre Řež, Husinec-Řež, Czech Republic) / Rojina Shrestha (Technical University Liberec, Liberec, Czech Republic) / Alena Ševců (Technical University Liberec, Liberec, Czech Republic)	118467
WP10.12	Evaluating the efficacy of peracetic formulations with CS by using electrochemical techniques	Yolanda De-Abreu (Nalco Champion, An Ecolab Company, Houston, United States of America) / Ramakrishnan Balasubramanian (Nalco Champion, An Ecolab Company, Houston, United States of America) / Alicia Jones (Nalco Champion, An Ecolab Company, Houston, United States of America) / Yingrui Zhang (Nalco Champion, An Ecolab Company, Houston, United States of America)	119920
WP10.13	Bacterial succession over 316L galvanized steel in marine environment	Artur Capao (Estacio University, Rio de Janeiro, Brazil) / Suleima Bitati (Estacio University, Rio de Janeiro, Brazil) / Luciano Procópio (Estacio University, Rio de Janeiro, Brazil)	120322

Corrosion of Steel in Concrete (WP11)

Posternumber	Paper title	Authors	Submission ID
WP11.1	Chloride Threshold of Prestressing Steel in Cementitious Systems with Ordinary Portland Cement, Fly Ash, and Corrosion Inhibitors	Dyana Joseline (Indian Institute of Technology, Chennai, India) / Peerzada Abdul Basit (Indian Institute of Technology Madras, Chennai, India) / Radhakrishna G Pillai (Indian Institute of Technology Madras, Chennai, India)	101447

Posternumber	Paper title	Authors	Submission ID
WP11.2	Corrosion Inhibition Effect of Polyvinylpyrrolidone on Reinforcing Steel in Simulated Concrete Pore Solutions	Ronggui Du (College of Chemistry and Chemical Engineering, Xiamen University, Xiamen, China PR) / Yingbo Gao (College of Chemistry and Chemical Engineering, Xiamen University, Xiamen, China PR) / Shigang Dong (College of Chemistry and Chemical Engineering, Xiamen University, Xiamen, China PR) / Zichao Guan (College of Chemistry and Chemical Engineering, Xiamen University, Xiamen, China PR) / Xia Wang (College of Chemistry and Chemical Engineering, Xiamen University, Xiamen, China PR)	102392
WP11.3	A new method to assess the corrosion rate of reinforced concrete structure	Gabriel Samson (LMDC, Université de Toulouse, INSA, UPS, Toulouse, France) / Fabrice Deby (LMDC, Université de Toulouse, INSA, UPS, Toulouse, France) / Jean-Luc Garciaz (LERM SETEC, Arles, France) / Jean-Louis Perrin (LERM SETEC, Arles, France)	107117
WP11.4	Corrosion behavior of iron in cementitious solution at 80°C in anoxic condition	David PALLY (Den-Service d'Etude du Comportement des Radionucléides (SECR), CEA, Université Paris-Saclay, F-91191, , Gif-sur-Yvette, France) / Bernard Grenut (Den-Service d'Etude du Comportement des Radionucléides (SECR), CEA, Université Paris-Saclay, F-91191,, Gif-sur-Yvette, , France) / Frederic MISERQUE (Den-Service de la Corrosion et du Comportement des Matériaux dans leur environnement (SCCME), CEA, Université Paris-Saclay, F-91191, , Gif-sur-Yvette, , France) / Patrick Le BESCOP (Den-Service d'Etude du Comportement des Radionucléides (SECR), CEA, Université Paris-Saclay, F-91191, , Gif-sur-Yvette, , France) / Laure CHOMAT (Den-Service d'Etude du Comportement des Radionucléides (SECR), CEA, Université Paris-Saclay, F-91191, , Gif-sur-Yvette, , France) / Valérie L'HOSTIS (Den-Service d'Etude du Comportement des Radionucléides (SECR), CEA, Université Paris-Saclay, F-91191, , Gif-sur-Yvette, , France)	108312
WP11.5	Impact of low alkali cement in ordinary concrete on ingress of chloride ions	Zofia Szweda (Silesian University of Technology, Gliwice, Poland)	111937
WP11.6	Inhibition Efficiency of biopolymer against Carbon Steel Corrosion in simulated pores concrete solution	Manoel Martins (IFAL, Maceió, Brazil) / Josealdo Tonholo (UFAL, Maceió, Brazil) / Silvia B. B. Uchoa (UFAL, Maceió, Brazil) / Davi Vieira (IFAL, Maceió, Brazil)	113352

Posternumber	Paper title	Authors	Submission ID
WP11.7	Migration and corrosion protection tests of cationic corrosion inhibitors	Tomas Lovasi (University of Chemistry and Technology, Prague, Czech Republic) / Milan Kouril (University of Chemistry and Technology, Prague, Czech Republic) / Sarka Msallamova (University of Chemistry and Technology, Prague, Czech Republic) / Benazir Jursunbek (University of Chemistry and Technology, Prague, Czech Republic) / Vendula Simunkova (student, Prague, Czech Republic)	113807
WP11.8	Corrosion Testing on Steel Reinforced XD3 Concrete Samples Prepared with a Green Inhibitor and two Different Superplasticizers	Shaymaa Abbas Abdulsada (Faculty of Materials Science and Engineering, University of Miskolc, Hungary, Miskolc, Hungary) / Éva Fazakas (Department of Surface Technology, Bay Zoltán Nonprofit Ltd. for Applied Research, Budapest, Hungary) / Tamás I. Török (Faculty of Materials Science and Engineering, University of Miskolc, Miskolc, Hungary)	117727
WP11.9	Effects of Oxygen Partial Pressure on Corrosion Behavior of P110 Steel in Nitrogen injection Wellbore	Yuanhua Lin (Southwest Petroleum University, Chengdu, China PR) / Zhi Ming Yu (Southwest Petroleum University, Chengdu, China PR) / Dezhi Zeng (Southwest Petroleum University, Chengdu, China PR)	118787
Corrosion in Oil and Gas Production (WP13)			
WP13.11	Pitting corrosion in low carbon steels exposed to Mexican soil: nucleation and growth	Antonio Contreras (Mexican institute of petroleum, Mexico, Mexico) / Luis Quej (Mexican institute of petroleum, Mexico, Mexico) / Hongbo Liu (Mexican institute of Petroleum, Mexico, Mexico) / Eliceo Sosa (Mexican institute of petroleum, Mexico, Mexico) / Jorge Alamilla (Mexican institute of petroleum, Mexico, Mexico)	121232
WP13.10	An impending inhibitor useful for the petroleum industry	Kashif Rahmani Ansari (Centre of Research Excellence in Corrosion, Research Institute, King Fahd University of Petroleum and Minerals, Dhahran 31261, Saudi Arabia, Dhahran, Saudi Arabia) / M. A. Quraishi (Centre of Research Excellence in Corrosion, Research Institute, King Fahd University of Petroleum and Minerals, Dhahran 31261, Saudi Arabia, Dhahran , Saudi Arabia)	120342
WP13.9	Stainless Steel Amine Reboiler corrosion assessment	Cinzia Foschini (ENI SPA, San donato Milanese, Italy) / Giacomo Di Jorio (ENI SPA, San donato milanese, Italy) / Giacomo Garzia (ENI SPA, San donato milanese, Italy) / Salvatore Montalto (ENI SPA, San Donato MIlanese, Italy) / Pietro Ciccorossi (ENI SPA, San Donato Milanese, Italy)	119082

Posternumber	Paper title	Authors	Submission ID
WP13.7	Enabling remote corrosion detection on Oil and Gas installations with the stand-alone Field Kelvin Probe (FKP)	Eugen Florin Turcu (Christian Michelsen Research, Bergen, Norway) / Bård Henriksen (Christian Michelsen Research, Bergen, Norway) / Inge Klepsvik (Christian Michelsen Research, Bergen, Norway) / Gaute Øverås Lied (Christian Michelsen Research, Bergen, Norway)	111617
WP13.6	Sulfide Stress Cracking and Hydrogen Induced Cracking Susceptibility of High Strength Carbon Steel Wires	Pedro Netto da Silva (Federal University of Rio de Janeiro, Rio de Janeiro, Brazil) / Emanuel Seixas Nascimento Filho (Federal University of Rio de Janeiro, Rio de Janeiro, Brazil) / Jonas da Silva de Sá (Federal University of Rio de Janeiro, Rio de Janeiro, Brazil) / Eduardo Alencar de Souza (Federal University of Rio de Janeiro, Rio de Janeiro, Brazil) / José Antônio da Cunha Ponciano Gomes (Federal University of Rio de Janeiro, Rio de Janeiro, Brazil)	110657
WP13.5	Influence on the sulfide-containing films content on the corrosion and hydrogenation particularities of steels of various structures in chloride-acetate solution	Myroslav Khoma (National Academy of Scienses of Ukraine, Karpenko Physico-Mechanical Institute, Lviv, Ukraine) / Svitlana Golovey (National Academy of Scienses of Ukraine, Karpenko Physico-Mechanical Institute, Lviv, Ukraine) / Vasyl Ivashkiv (National Academy of Scienses of Ukraine, Karpenko Physico-Mechanical Institute, Lviv, Ukraine) / Chrystyna Vasyliv (National Academy of Sciences of Ukraine Karpenko Physico-Mechanical Institute, Lviv, Ukraine) / Marian Chuchman (National Academy of Sciences of Ukraine Karpenko Physico-Mechanical Institute, Lviv, Ukraine) / Nadiia Ratska (National Academy of Sciences of Ukraine Karpenko Physico-Mechanical Institute, Lviv, Ukraine)	108352
WP13.4	The influence of Cr on the electrochemical corrosion and scale formation behaviors of high Mn steel in a sour environment	Sung Jin Kim (Sunchon National University, Sunchon, Korea, South) / Jin Ho Park (POSCO, Pohang, Korea, South) / Eun Hye Hwang (Sunchon National University, Sunchon, Korea, South) / Ho Jong Lee (Sunchon National University, Sunchon, Korea, South)	107347
WP13.3	Effect of extremely aggressive environment on the nature of passive film of HP-13Cr stainless steel	Xuanpeng Li (Harbin Engineering University, Harbin, China PR) / Tao Zhang (Harbin Engineering University, Harbin, China PR) / Fuhui Wang (Northeastern University, Shenyang, China PR)	106332
WP13.2	Role of inclusions in localized corrosion initiation of X70 steel in near-neutral pH environment	Liwei Wang (Qingdao University, Qingdao, China PR)	104682

Posternumber	Paper title	Authors	Submission ID
WP13.1	Corrosion behavior of HP-13Cr stainless steel in the extremely aggressive oilfield environment: Part II. Effect of acid stimulation	Yang Zhao (Corrosion and protection Laboratory, Key Laboratory of Superlight Materials and Surface Technology, Harbin Engineering University, Ministry of Education, Nantong ST 145, Harbin, China PR) / Junfeng Xie (Petrochina Tarim Oilfield Company, Shihua RD, Korla, China PR) / Guanxin Zeng (Petrochina Tarim Oilfield Company, Shihua RD, Korla, China PR) / Tao Zhang (Key Laboratory for Anisotropy and Texture of Materials, School of Materials Science and Engineering, Northeastern University, Wenhua Road 3-11, Shenyang, China PR) / Dake Xu (Key Laboratory for Anisotropy and Texture of Materials, School of Materials Science and Engineering, Northeastern University, Wenhua Road 3-11, Shenyang , China PR) / Fuhui Wang (Key Laboratory for Anisotropy and Texture of Materials, School of Materials Science and Engineering, Northeastern University, Wenhua Road 3-11, Shenyang, China PR)	101311
Inorganic Coatings			
WP14.IC.1	PEO coatings on AZ31 magnesium alloy with incorporated particles: wear and corrosion behavior	Marta Mohedano (Universidad Complutense de Madrid, Madrid, Spain) / Marina Corbacho (Universidad Complutense de Madrid, Madrid, Spain) / Raul Arrabal (Universidad Complutense de Madrid, Madrid, Spain) / Beatriz Mingo (Helmholtz Zentrum Geesthacht, Geesthacht) / Angel Pardo (Universidad Complutense de Madrid, Madrid, Spain) / Endzhe Matykina (Universidad Complutense de Madrid, Madrid, Spain)	105557
WP14.IC.2	Electrochemical behaviour of aluminium alloys in KOH-based electrolytes	Delphine VEYS-RENAUX (Université de Lorraine, Vandoeuvre les Nancy, France) / Najat Chahboun (Université de Lorraine, Vandoeuvre les Nancy, France) / Mohamed Moalla (Ecole Nationale Supérieure d'Ingénieurs de Sfax, Sfax, Tunisia) / Anas Ben Romdhane (Ecole Nationale Supérieure d'Ingénieurs de Sfax, Sfax, Tunisia) / Khaled Elleuch (Ecole Nationale Supérieure d'Ingénieurs de Sfax, Sfax, Tunisia) / Emmanuel Rocca (Université de Lorraine, Vandoeuvre les Nancy, France)	108102
WP14.IC.3	High Frequency Pulse Anodising of Die Cast Aluminium Alloys	Flemming Jensen (Technical University of Denmark, Kgs. Lyngby, Denmark) / Visweswara Gudla (Technical University of Denmark, Kgs. Lyngby, Denmark) / Rajan Ambat (Technical University of Denmark, Kgs. Lyngby, Denmark) / Ib Kongstad (Bang & Olufsen A/S, Struer, Denmark)	110147

Posternumber	Paper title	Authors	Submission ID
WP14.IC.4	Determination of processing parameters suitable for "soft sparking" during plasma electrolytic oxidation of aluminium	Maciej Sowa (Silesian University of Technology, Gliwice, Poland) / Kamil Boroń (Silesian University of Technology, Gliwice, Poland) / Artur Maciej (Silesian University of Technology, Gliwice, Poland) / Alicja Kazek-Kęsik (Silesian University of Technology, Gliwice, Poland) / Wojciech Simka (Silesian University of Technology, Gliwice, Poland)	110837
WP14.IC.5	Investigation of DLC coated steel surfaces corrosion resistance: an electrochemical imaging approach	Thibault MAERTEN (OERLIKON Balzers, LIMOGES, France) / Roland OLTRA (ICB, UMR CNRS 6303, Université Bourgogne Franche Comté, Dijon, France) / Cédric JAOUL (IRCER, UMR CNRS 7315, Université de Limoges, Limoges, France) / Christophe LE NINIVEN (IRCER, UMR CNRS 7315, Université de Limoges, Limoges, France) / Pascal TRISTANT (IRCER, UMR CNRS 7315, Université de Limoges, Limoges, France) / Frédéric MEUNIER (OERLIKON Balzers, Limoges, France) / Olivier JARRY (OERLIKON Balzers, Balzers, Liechtenstein)	111647
WP14.IC.6	Effect of Cu-content on microstructure and corrosion behavior of PEO coating on Al–xCu alloys	Liye Zhu (Harbin Engineering University, Harbin , China PR) / Tao Zhang (Harbin Engineering University, Harbin, China PR) / Wei Zhang (Institute of Metal Research, Chinese Academy of Sciences, Shenyang, China PR) / Fuhui Wang (Northeastern University, Shenyang, China PR)	113842
WP14.IC.7	Preparation and characterization of hydroxides-based composite film on magnesium alloys by steam coating	Takahiro Ishizaki (Shibaura Institute of Technology, Tokyo, Japan) / Yuta Shimada (Shibaura Institute of Technology, Tokyo, Japan) / Tomohiro Miyashita (Shibaura Institute of Technology, Tokyo, Japan) / Ai Serizawa (Shibaura Institute of Technology, Tokyo, Japan)	114327
WP14.IC.8	Preparation of Anticorrosive Film for Inhibiting Pitting Corrosion on Aluminum Alloy by Steam Coating	Ai Serizawa (Shibaura Institute of Technology, Tokyo, Japan) / Kohei Watanabe (Shibaura Institute of Technology, Tokyo, Japan) / Takahiro Ishizaki (Shibaura Institute of Technology, Tokyo, Japan)	115667
WP14.IC.9	Coatings obtained on titanium by Plasma Electrochemical Oxidation	Krzysztof Rokosz (Koszalin University of Technology, Koszalin, Poland) / Tadeusz Hryniewicz (Koszalin University of Technology, Koszalin , Poland) / Steinar Raaen (Norwegian University of Science and Technology (NTNU), Trondheim , Norway) / Sofia Gaiaschi (HORIBA Jobin Yvon S.A.S., Palaiseau, France) / Patrick Chapon (HORIBA Jobin Yvon S.A.S., Palaiseau, France) / Winfried Malorny (4Hochschule Wismar-University of Applied Sciences Technology, Wismar) / Lukasz Dudek (Koszalin University of Technology, Koszalin, Poland) / Kornel Pietrzak (Koszalin University of Technology, Koszalin, Poland) / João Carlos Salvador Fernandes (Instituto Superior Técnico University of Lisbon, Lisbon, Portugal)	117842

Posternumber	Paper title	Authors	Submission ID
WP14.IC.10	The influence of fiber laser radiation on the corrosion resistance of 7075 aluminium alloy	Katarzyna Lecka (Wroclaw University of Science and Technology, Wroclaw, Poland) / Arkadiusz Antonczak (Wroclaw University of Science and Technology, Wroclaw, Poland)	118672
WP14.IC.12	Corrosion resistance of stainless steel covered by phosphomolybdate ion encapsulated in silica	LIDIA ADAMCZYK (Częstochowa Technical of University, Częstochowa, Poland)	120122
WP14.IC.13	Characterization of plasma electrolytic oxidation influence on corrosion behaviour of AZ31 magnesium alloy in various chloride containing environments	Daniel Kajánek (University of Zilina, Zilina, Slovakia) / Filip Pastorek (Research Centre, University of Zilina, Zilina, Slovakia) / Branislav Hadzima (Research Centre, University of Zilina, Zilina, Slovakia)	120792
WP14.IC.14	Microstructure and corrosion behaviors of Hf/HfN multilayer coatings on magnesium alloys	Zhoucheng Wang (Xiamen University, Xiamen, China PR)	120972
Metallic Coatings			
	Advanced coating strategies to face industrial challenges of zinc coatings	Sudesh Wijesinghe (Singapore Institute of Manufacturing Technology (SIMTech), Singapore, Singapore) / Wenjin Yan (Singapore Institute of Manufacturing Technology (SIMTech), Singapore, Singapore) / Linda Y L Wu (Singapore Institute of Manufacturing Technology (SIMTech), Singapore, Singapore) / Yong Teck Tan (Singapore Institute of Manufacturing Technology (SIMTech), Singapore, Singapore) / Qian Min (Singapore Institute of Manufacturing Technology (SIMTech), Singapore, Singapore) / Chan Wai Koh (Singapore Institute of Manufacturing Technology (SIMTech), Singapore , Singapore)	101966
WP.14.MC.1	Study on application of Zn-Mg-Al coated steel sheet for logistic pallet	Jae-Won Lee (Pohang institute of metal industry advencement, KS010, Korea, South) / Kyung Seh Lee (Pohang Institute of metal Industry advancement, KS010, Korea, South) / Sungyoul Choi (2Samjung industries company, KS010, Korea, South)	103492
WP.14.MC.2	Ultrasound assisted electroless deposition of nickel-boron coatings: Effect of ultrasonic frequency on the tribocorrosion behavior	Véronique Vitry (UMONS, MONS, Belgium) / Luiza Bonin (UMONS, MONS, Belgium)	107647

Posternumber	Paper title	Authors	Submission ID
WP.14.MC.3	The effect of Mg concentration on the microstructure and corrosion resistance of ZnxMg1-x alloy coatings	KYOUNG PIL KO (POSCO, Gwangyang-si, Korea, South) / Kyung-Hoon Nam (POSCO Technical Research Laboratory, Gwangyang-si, Korea, South) / Yong-Hwa Jung (POSCO Technical Research Laboratory , Gwangyang-si, Korea, South) / Dong-Yeoul Lee (POSCO Technical Research Laboratory , Gwangyang-si, Korea, South) / Moon-Jong Eum (POSCO Technical Research Laboratory , Gwangyang-si, Korea, South) / Young-Jin Kwak (POSCO Technical Research Laboratory , Gwangyang-si, Korea, South) / Seok-Joon Hong (POSCO Technical Research Laboratory , Gwangyang-si, Korea, South) / Tae-Yeob Kim (POSCO Technical Research Laboratory , Gwangyang-si, Korea, South)	108207
WP.14.MC.4	Corrosion resistance of Zn-Ni-Al2O3 nanocomposite coatings	Gabriella Roventi (Università Politecnica delle Marche, Ancona, Italy) / Giampaolo Giuliani (Università Politecnica delle Marche, Ancona, Italy) / Annamaria Viceré (Università Politecnica delle Marche, Ancona, Italy) / Tiziano Bellezze (Università Politecnica delle Marche, Ancona, Italy)	109102
WP.14.MC.5	Comparison of carbide and alloy based HVOF sprayed coatings using salt spray testing	Jan Schubert (Vyzkumny a zkusebni ustav Plzen s.r.o., Pilsen, Czech Republic) / Zdenek Cesanek (Vyzkumny a zkusebni ustav Plzen s.r.o., Plzen, Czech Republic) / Kateřina Kopelentová (Vyzkumny a zkusebni ustav Plzen s.r.o., Plzen, Czech Republic)	109882
WP.14.MC.6	Numerical Modelling and Impact of Cu addition on Microstructure and Surface Characteristics of Laser Deposited Ti-6Al-4V Alloy.	OLAWALE FATOBA (UNIVERSITY OF JOHANNESBURG, JOHANNESBURG, South Africa) / MAMOOKHO MAKHATHA (UNIVERSITY OF JOHANNESBURG, JOHANNESBURG, South Africa) / ESTHER AKINLABI (UNIVERSITY OF JOHANNESBURG, JOHANNESBURG, South Africa)	111497
WP.14.MC.7	INFLUENCE OF THE SYNTHESIS PARAMETERS VARIATION ON THE HYDROPHOBICITY PROPERTIES OF CuO COATINGS	Jennifer Johanna González Motta (Industrial University of Santander, Floridablanca , Colombia) / Maria Carolina Florez Ariza (Industrial University of Santander, Bucaramanga, Colombia) / Sergio Ismael Blanco Vásquez (Industrial University of Santander, Bucaramanga, Colombia)	113487
WP.14.MC.8	The effect of cut edges on HDG coating quality	Hana Geiplova (SVÚOM Ltd., Prague, Czech Republic) / Antonin Schwarzer (Signum Ltd., Hustopece, Czech Republic) / Zdenek Bartak (SVÚOM Ltd., Prague, Czech Republic)	113707
WP.14.MC.9	Properties of HDG coatings	Zdenek Bartak (SVÚOM Ltd., Prague, Czech Republic) / Katerina Kreislova (SVÚOM Ltd., Prague, Czech Republic) / Antonin Schwarzer (Signum Ltd., Hustopece, Czech Republic)	113752

Posternumber	Paper title	Authors	Submission ID
WP.14.MC.10	Design of experiment of microsized and nanostructured HVOF WC-12Co coating HVOF for Erosion Performance	Bander Aldaajani (Dublin City University, Dublin, Ireland) / Sultan AlMutairi (Saudi Aramco, Dhahran, Saudi Arabia) / Khaled Benyounis (DCU, Dublin, Ireland) / Joseph Stoles (DCU, Dublin, Ireland) / Ahmed Alhamed (DCU, Dublin, Ireland)	115782
WP.14.MC.11	Electrodeposition of Ni/cerium molybdate composite coatings from deep eutectic solvent	Juliusz Winiarski (Wroclaw University of Science and Technology, Wroclaw, Poland) / Beata Cieślikowska (Wroclaw University of Science and Technology, Wrocław, Poland) / Anna Mazur (Wroclaw University of Science and Technology, Wrocław, Poland) / Bogdan Szczygieł (Wroclaw University of Science and Technology, Wrocław, Poland)	117632
WP.14.MC.12	Corrosion Resistance of Cold Sprayed Nickel Coatings in Acidic Chloride Solution	Mieczyslaw Scendo (Jan Kochanowski University in Kielce, Kielce, Poland) / Wojciech Zorawski (University of Technology in Kielce, Kielce, Poland) / Katarzyna Staszewska-Samson (Jan Kochanowski University in Kielce, Kielce, Poland) / Klaudia Szczerba (Jan Kochanowski University in Kielce, Kielce, Poland)	117702
WP.14.MC.13	Effect of magnetron sputtering of active powder with MnCrFeNiCo layers on electrochemical parameters of metal hydride electrode	AGNIESZKA STEFANIAK (Czestochowa University of Technology, Czestochowa Czestochowa, Poland) / Henryk Bala (Czestochowa University of Technology, Czestochowa, Poland)	118417
WP.14.MC.14	Preparation of thin Ni coatings on active hydride powder material using magnetron sputtering technique	Klaudia Bordolinska (Czestochowa University of Technology, Czestochowa, Poland) / Henryk Bala (Czestochowa University of Technology, Czestochowa, Poland)	118587
WP.14.MC.15	Interdiffusion Coatings for High-Temperature-Corrosion Protection of Low-Alloy Steels	Kamil Balinski (University of Applied Sciences Osnabrück, Osnabrück) / Katrin Jahns (University of Applied Sciences Osnabrück, Osnabrück) / Manuel Schroth (University of Applied Sciences Osnabrück, Osnabrück) / Sue Impey (Cranfield University, Cranfield, United Kingdom) / Christine Chalk (Cranfield University, Cranfield, United Kingdom) / Jürgen Wübbelmann (University of Applied Sciences Osnabrück, Osnabrück) / John Nicholls (Cranfield University, Cranfield, United Kingdom) / Ulrich Krupp (University of Applied Sciences Osnabrück, Osnabrück)	119262
WP.14.MC.16	Effects of Dilution Rate and Parameter Sensitivity on the Microstructure and Mechanical Properties of Laser Cladded Ti-Cu-Al Coatings on Ti-6Al-4V Alloy.	OLAWALE FATOBA (UNIVERSITY OF JOHANNESBURG, JOHANNESBURG, South Africa) / STEPHEN AKINLABI (UNIVERSITY OF JOHANNESBURG, JOHANNESBURG, South Africa) / ESTHER AKINLABI (UNIVERSITY OF JOHANNESBURG, JOHANNESBURG, South Africa)	120047

Posternumber	Paper title	Authors	Submission ID
WP.14.MC.17	Surface analysis of Inconel CMT coatings after the corrosion in the renewable fuel ash environments	Aleksandra Dębowska (AGH University of Science and Technology, Cracow, Poland) / Agnieszka Kopia (AGH University of Science and Technology, Cracow, Poland) / Aneta Magdziarz (AGH University of Science and Technology, Cracow, Poland)	120637
WP.14.MC.18	Microstructure and performance of external protection coating on ductile iron pipes	Anna Wassilkowska (Cracow University of Technology, Kraków, Poland) / Agnieszka Kochmańska (West Pomeranian University of Technology Szczecin, Szczecin, Poland)	120700
WP.14.MC.19	A comparative study on the corrosion behaviors of metallic coatings deposited on AZ91D magnesium alloys	Zhoucheng Wang (College of Chemistry and Chemical Engineering, Xiamen University, Xiamen, China PR)	120982

Organic Coatings

WP.14.OC.2	The graphene oxide–poly(urea–formaldehyde) composites used in corrosion protection on mild steel	Hongpeng ZHENG (Harbin Engineering University, Harbin Harbin, China PR) / Yawei SHAO (Harbin Engineering University, Harbin, China PR)	102076
WP.14.OC.3	Failure process of modified iron oxide/epoxy coating under alternate wetting and drying condition	guirong wang (Harbin Engineering University, Harbin, China PR) / yawei shao (Harbin Engineering University, Harbin , China PR)	103112
WP.14.OC.5	MATChING project: Coatings development for corrosion protection in low-T geothermal power plants	Lorena Freire Pineiro (AIMEN Technology Center, O Porrino, Spain) / Alberto Fernández (AIMEN Technology Center, O Porrino, Spain) / Ricardo Losada (Danish	108067
WP.14.OC.6	EVALUATING THE INFLUENCE OF SURFACE PREPARATION ON THE PERFORMANCE OF EPOXY PRIMERS SYSTEM PIGMENTED WITH ZINC PHOSPHATE	Idalba Souza dos Santos (Brazilian Navy, Niterói, Brazil) / Ladimir José de Carvalho (Federal University of Rio de Janeiro, EQ/DPI, Rio de Janeiro, Brazil) / Simone Louise Delarue Cezar Brasil (Federal University of Rio de Janeiro, EQ/DPI, Rio de Janeiro, Brazil)	116872
WP.14.OC.7	Influence of humidity on the electrochemical evaluation method of organic coating containing glass flakes in real environment	Daisuke Ito (Yokohama National University, Yokohama, Japan) / Kouichi Sekino (Kanagawa Institute of Industrial Science and Technology, Ebina, Japan) / Shinji Okazaki (Yokohama National University, Yokohama, Japan) / Kazuyoshi Sekine (High Pressure Institute of Japan, Tokyo, Japan) / Masaru Ishihara (Japan Oil, Gas and Metals National Corporation, Tokyo, Japan)	118892
WP.14.OC.8	Anticorrosive polyurethane coatings for renovation of the widebody aircrafts shell.	Barbara Pilch-Pitera (Rzeszow University of Technology, Rzeszow, Poland) / Dominika Czachor (Rzeszow University of Technology, Rzeszow, Poland) / Łukasz Byczyński (Rzeszow University of Technology, Rzeszow, Poland) / Joanna Wojturska (Rzeszow University of Technology, Rzeszow, Poland) / Robert Ostatek (Rzeszow University of Technology, Rzeszow, Poland)	119427

Posternumber	Paper title	Authors	Submission ID
WP.14.OC.9	Pinholes on coating applied over TSZ	Lukasz Augustynski (Rosa , Warszawa, Poland) / Krzysztof Szymanski (Politechnika Śląska, Katowice, Poland)	119682
WP.14.OC.10	Biocompatibility of Electrochemically Reduced Graphene Oxide on biomedical grade CoCr alloy. In vivo response of graphene	Maria Lorenza Escudero (CENIM-CSIC, Madrid, Spain) / Irene Llorente (CENIM-CSIC, Madrid, Spain) / Blanca Teresa Perez-Maceda (CIB-CSIC, Madrid, Spain) / Sara San José Pinilla (CIB-CSIC, Madrid, Spain) / Rosa Maria Lozano (CIB-CSIC, Madrid, Spain) / Soledad Aguado Henche (University of Alcala de Henares, Madrid, Spain) / Celia Clemente de Arriba (University of Alcala de Henares, Madrid, Spain) / Miguel Angel Alobera (University of León, León, Spain) / Maria Cristina Garcia-Alonso (CENIM-CSIC, Madrid, Spain)	120992
WP.14.OC.11	The influence of the concentration and shape, size and surface modification of Zn pigments on the corrosion behaviour of coatings	Ewa Langer (Institute for Engineering of Polymer Materials and Dyes, Paint and Plastics department, Gliwice, Poland) / Małgorzata Zubielewicz (Institute for Engineering of Polymer Materials and Dyes, Paint and Plastics department , Gliwice, Poland) / Helena Kuczyńska (Institute for Engineering of Polymer Materials and Dyes, Paint and Plastics department , Gliwice, Poland) / Agnieszka Królikowska (Road and Bridge Research Institute, Warsaw, Poland) / Leszek Komorowski (Road and Bridge Research Institute , Warsaw, Poland)	121740

Pretreatments

Posternumber	Paper title	Authors	Submission ID
WP.14.PT.1	Characteristics of TBOT-MAPTS organic-inorganic hybrid coatings on stainless steel	Wan Young Maeng (Korea Atomic Energy Research Institute, Daejeon, Korea, South) / Dong Jin Kim (Korea Atomic Energy Research Institute, Daejeon, Korea, South)	103692
WP.14.PT.2	Effect of pre-treatment on corrosion resistance of phosphate conversion coating on Mg alloy	Yuyang Chen (Harbin Engineering University, Harbin, China PR) / Tao Zhang (Northeastern University, Shenyang, China PR) / Fuhui Wang (Northeastern University, Shenyang, China PR)	103742
WP.14.PT.3	Effect of mechanical surface pre-treatment on the microstructure and corrosion resistance of the phosphate conversion coating of AZ91D magnesium alloy	Chunyan Zhang (Harbin Engineering University, Harbin, China PR) / Tao Zhang (Harbin Engineering University, Harbin, China PR) / Baoxing Yu (Institute of Metal Research, Chinese Academy of Sciences, Shenyang, China PR) / Fuhui Wang (Northeastern University, Shenyang, China PR)	106202

Posternumber	Paper title	Authors	Submission ID
WP.14.PT.4	Influence of an aging time and the stabilization process of the organic-inorganic sol-gel coating on corrosion mitigation	Jolanta Gąsiorek (Department of Mechanics, Materials Science and Engineering, Wroclaw University of Science and Technology, Smoluchowskiego 25, 50-370 Wrocław, Wroclaw, Poland) / Justyna Krzak (Department of Mechanics, Materials Science and Engineering, Wroclaw University of Science and Technology, Smoluchowskiego 25, 50-370 Wrocław, Wroclaw, Poland) / Jerzy Kaleta (Department of Mechanics, Materials Science and Engineering, Wroclaw University of Science and Technology, Smoluchowskiego 25, 50-370 Wrocław, Wroclaw, Poland)	107627
WP.14.PT.5	Investigation of zirconium based conversion coatings applied on aluminium alloy ENAC-AlSi7Mg0.3	Gavrilo Šekularac (Institute Jožef Stefan, Ljubljana, Slovenia) / Ingrid Milošev (Institute Jožef Stefan, Ljubljana, Slovenia)	110112
Self-healing and smart Coatings			
WP.14.SH.1	The self-healing effect of hybrid sol-gel coatings based on CeO2 nanoparticles modified with Ce3+ ions applied on AA7075-T6	Urša Tiringer (Institut Jožef Stefan, Ljubljana, Slovenia) / Gavrilo Šekularac (Institut Jožef Stefan, Ljubljana, Slovenia) / Janez Kovač (Institut Jožef Stefan, Ljubljana, Slovenia) / Ingrid Milošev (Institut Jožef Stefan, Ljubljana, Slovenia)	109962
Corrosion in Refinery and Petrochemistry (WP15)			
WP15.3	Sulphidic corrosion study of AISI-SAE 1020 steel exposed to Dimethyl Sulfide in a simulated environment	Dionisio Antonio Laverde Cataño (Corrosion Investigation Group, Universidad Industrial de Santander, Bucaramanga, Colombia) / Javier Alberto Sanabria Cala (Corrosion Investigation Group, Universidad Industrial de Santander, Bucaramanga, Colombia) / Gerson Rafael Conde Rodríguez (Corrosion Investigation Group, Universidad Industrial de Santander, Bucaramanga, Colombia) / Dario Yesid Peña Ballesteros (Corrosion Investigation Group, Universidad Industrial de Santander, Bucaramanga, Colombia)	119410

Posternumber	Paper title	Authors	Submission ID
Cathodic Protection (WP16)			
WP16.3	Galvanic Corrosion Behavior of Bolt and Nut Combinations	Shiyan Zhang (Southwest Technology and Engineering Research Institute, Chongqing, China PR) / Xiaokui Yang (Southwest Technology and Engineering Research Institute, Chongqing, China PR) / Wanjun Yang (Southwest Technology and Engineering Research Institute, Chongqing, China PR) / Xiaoqin Wei (Southwest Technology and Engineering Research Institute, Chongqing, China PR) / Bingfei Hu (Southwest Technology and Engineering Research Institute, Chongqing, China PR) / Wei Liu (China South Industries Group Co., LTD., Beijing, China PR)	122135
WP16.1	Preparation of anatase-TiO2 thin films by micro-mixing process for visible light-induced photo-generated cathodic protection	ZUDE FENG (College of Materials, Xiamen University, Xiamen, China PR) / Chao Xu (College of Materials, Xiamen University, Xiamen, China PR)	101907
WP16.2	Effectiveness of different coatings on cathodic protection engineering for gas pipelines	Saeed Ahmadizadeh (National Iranian Gas Co. (NIGC), Rasht, Iran)	104862
Automotive Corrosion (WP17)			
WP17.4	Impact of surface coating on the corrosion resistance of the brazed condensor	Christian CASENAVE (VALEO Thermal System, La Verrière, France) / Maryse PHILIPPE (VALEO Thermal System, La Verrière, France) / Anne-Gaëlle Villemiane (VALEO Thermal System, La Verrière, France)	120132
WP17.1	In situ investigation of corrosion process of press hardened steel coated with hot-dip Al-Si and electroplated Zn-Ni using synchrotron diffraction	Camila Pucci Couto (IPEN - Nuclear and Energy Research Institute, SAO PAULO SAO PAULO, Brazil) / Isolda Costa (IPEN - Nuclear and Energy Research Institute , Sao Paulo, Brazil) / Luis Martinez (IPEN - Nuclear and Energy Research Institute, Sao Paulo, Brazil) / Geraldo Carezzato (IPEN -Nuclear and Energy Research Institute, Sao Paulo, Brazil) / Marco Colosio (GM - General Motors Mercosul, São Caetano do Sul, Brazil) / Xabier Turrilas (ICMAB/CSIC - Institute of Materials Science of Barcelona, Barcelona, Spain) / Zehbour Panossian (IPT – Institute for Technological Research, Sao Paulo, Brazil) / Jesualdo Rossi (IPEN - Nuclear and Energy Research Institute, Sao Paulo, Brazil) / Maurilio Gomes (IPEN - Nuclear and Energy Research Institute , São Paulo, Brazil)	102242

Posternumber	Paper title	Authors	Submission ID
WP17.2	Generalized interpretation of Hydrogen signal measured by Thermal Desorption Spectroscopy through newly-devised O2-purge/Ar-carrier gas chromatograph apparatus.	Yasuhide Ishiguro (JFE Steel Corporation, Handa, Japan) / Shinji Ootsuka (JFE Steel Corporation, Kawasaki, Japan)	106962
WP17.3	Corrosion resistance investigation of press-hardened steel 22MnB5 with electrodeposited Zn-Ni coating	Jesualdo Luiz Rossi (Instituto de Pesquisas Energéticas e Nucleares, São Paulo, Brazil) / Priscilla Danielle de Oliveira Lopes Costa (IPEN, São Paulo, Brazil) / Camila Pucci Couto (IPEN, São Paulo, Brazil) / Isolda Costa (IPEN, São Paulo, Brazil)	114757

Tribo-Corrosion (WP18)

Posternumber	Paper title	Authors	Submission ID
WP18.1	Tribocorrosion performance of Zn, Zn-Ni and Zn-Ni with magnesium electrodeposited on medium carbon steel in a chloride environment	Marcin Kowalski (Warsaw University of Technology, Faculty of Civil Engineering, Mechanics and Petrochemistry, Lukasiewicza 17, 09-400 Plock, Poland) / Arkadiusz Stachowiak (Poznan University of Technology, Institute of Machines and Motor Vehicles,, Piotrowo Street 3, 60-965 Poznan, Poland)	103227
WP18.2	Fretting and fretting-corrosion processes of Ti6Al4V alloys in simulated oral cavity environment	Marcin Klekotka (Bialystok University of Technology, Białystok, Poland) / Jan Ryszard Dąbrowski (Bialystok University of Technology, Białystok, Poland)	105992
WP18.4	Influence of Temperature on the Corrosion Behaviour of Stainless Steels under Tribological Stress	Matthias Sorg (Institute for Materials System Technology Thurgau, WITg, Tägerwilen, Switzerland) / Arnulf Hoertnagl (Institute for Materials System Technology Thurgau, WITg, Tägerwilen, Switzerland) / Paul Gümpel (Institute for Materials System Technology Thurgau, WITg, Tägerwilen, Switzerland) / Philipp Schuler (Institute for Materials System Technology Thurgau, WITg, Tägerwilen, Switzerland)	111107
WP18.5	Investigating the effect Nevamine-CP20 inhibitor on the electrochemical properties of lean duplex stainless steel under dynamic conditions	Ihsan Ul Haq Toor (KFUPM, Dhahran, Saudi Arabia) / Farid Abdallah (KFUPM, Dhahran, Saudi Arabia)	121332

Corrosion of Polymer and Advanced Materials (WP19)

Posternumber	Paper title	Authors	Submission ID
WP19.1	Investigations on the degree of damage of polyethylene grades of heating oil storage tanks after a service life of more than 30 years	Margit Weltschev (BAM, Berlin) / Gabi Hoffmann (BAM, Berlin) / Anka Kohl (BAM, Berlin)	121840

Posternumber	Paper title	Authors	Submission ID
Corrosion and Corrosion Protection of Drinking Water Systems (WP20)			
WP20.1	Characterization of corrosion deposits in water supply pipes using scanning electron microscopy	Anna Wassilkowska (Cracow University of Technology (CUT), Kraków, Poland) / Joanna Bak (Institute of Water Supply and Environmental Monitoring, CUT, Kraków, Poland) / Maciej Gruszka (MPWiK SA, Kraków, Poland)	120685
Corrosion of Archaeological and Historical Artefacts (WP21)			
WP21.1	Preparation of stable artificial copper patina	Pavol Rak (Univeristy of chemistry and technology Prague, Prague, Czech Republic) / Richard Bureš (University of chemistry and technology Prague, Prague, Czech Republic) / Jan Stoulil (University of chemistry and technology Prague, Prague, Czech Republic)	108492
WP21.2	Corrosion behaviour of different bronze after 15 years of atmospheric exposure	Zdenek Bartak (SVÚOM Ltd., Prague, Czech Republic) / Katerina Kreislova (SVÚOM Ltd., Prague, Czech Republic) / Pavlina Fialova (SVÚOM Ltd., Prague, Czech Republic) / Pavel Maresovsky (SVÚOM Ltd., Prague, Czech Republic)	113727
WP21.3	Protection of lead in acetic acid containing model depository environment by absorbents and destimulators	Sarka Msallamova (University of Chemistry and Tecnology Prague, Prague, Czech Republic) / Kateryna Popova (University of Chemistry and technology Prague, Prague, Czech Republic) / Milan Kouril (University of Chemistry and Technology prague, Prague, Czech Republic) / Kristyna Charlotte Strachotova (University of Chemistry and Technology Prague, Prague, Czech Republic) / Pavla Dvorakova (University of Chemistry and Technology, Prague, Prague, Czech Republic)	115227
WP21.4	Corrosion protection of metallic substrates of historical interest by lanthanum-silica sol-gel coatings	F.R. García-Galván (National Center for Metallurgical Research/CENIM-CSIC, Madrid, Spain) / J. Pena-Poza (Institute of History/IH-CSIC, Madrid, Spain) / F. Agua (Institute of History/IH-CSIC, Madrid, Spain) / M.A. Villegas (Institute of History/IH-CSIC, Madrid, Spain) / J.C. Galván (National Center for Metallurgical Research/CENIM-CSIC, Madrid, Spain) / M. García-Heras (Institute of History/IH-CSIC, Madrid, Spain)	103357

Posternumber	Paper title	Authors	Submission ID
Corrosion Control in Aerospace (WP22)			
WP22.1	Effect of graphene oxide on anticorrosion performance of polyelectrolyte multilayer for 2A12 aluminum alloy substrates	Zhao Xia (Institute of Oceanology, Chinese Academy of Science, Qingdao, China PR) / Baorong Hou (Institute of Oceanology, Chinese Acedemy of Science, Qingdao, China PR) / Hongling Li (Institute of Oceanology, Chinese Academy of Science, Qingdao, China PR)	103617
WP22.3	Sulfuric acid anodization of Al and AA2024 with barrier reanodization	Tobias Leidens (UFRGS, Porto Alegre, Brazil) / Luis F. P. Dick (UFRGS, Porto Alegre, Brazil) / Gerhard Knörnschild (UFRGS, Porto Alegre, Brazil)	109737
Atmospheric corrosion (TF)			
TF1.1	Stable artificial copper alloys patinas prepared from gaseous phase	Richard Bures (UCT Prague, Prague 6 – Dejvice, Czech Republic) / Jan Stoulil (UCT Prague, Prague, Czech Republic) / Pavol Rak (UCT Prague, Prague, Czech Republic)	102777
TF1.2	Slow strain rate testing of the effect of atomic hydrogen in advanced high strength steels under atmospheric corrosion conditions	Radovan Goliáš (University of Chemistry and Technology Prague, Prague, Czech Republic) / Darya Rudomilova (University of Chemistry and Technology Prague, Technopark Kralupy, Kralupy nad Vltavou, Czech Republic) / Tomáš Prošek (University of Chemistry and Technology Prague, Technopark Kralupy, Kralupy nad Vltavou, Czech Republic) / Andreas Muhr (voestalpine Stahl GmbH, Linz, Austria) / Gerald Luckeneder (voestalpine Stahl GmbH, Linz, Austria) / Hubert Duchaczek (voestalpine Stahl GmbH, Linz, Austria)	109572
TF1.3	Atmospheric corrosion of structure metals in indoor environments with low corrosivity	Katerina Kreislova (SVUOM Ltd., Prague, Czech Republic) / Hana Geiplova (SVUOM Ltd., Prague, Czech Republic)	113637
TF1.4	Field Kelvin Probe – aided condition evaluation and prediction for coated / uncoated metallic surfaces exposed to atmosphere	Eugen Florin Turcu (Christian Michelsen Research, Bergen, Norway) / Bård Henriksen (Christian Michelsen Research, Bergen, Norway) / Inge Klepsvik (Christian Michelsen Research, Bergen, Norway) / Gaute Øverås Lied (Christian Michelsen Research, Bergen, Norway)	118932

Posternumber	Paper title	Authors	Submission ID
	A simple two-step method for fabrication of rare-earth-containing superhydrophobic protective film with good environmental adaptability	Jianxin He (Southwest Technology and Engineering Research Institute, Chongqing, China PR) / Lunwu Zhang (Southwest Technology and Engineering Research Institute, Chongqing, China PR) / Yong Luo (Southwest Technology and Engineering Research Institute, Chongqing, China PR) / Lei Zhu (Southwest Technology and Engineering Research Institute, Chongqing, China PR) / Wanjun Yang (Southwest Technology and Engineering Research Institute, Chongqing, China PR) / Lei Liu (School of Chemistry and Chemical Engineering, Chongqing University, Chongqing, China PR) / Lingjie Li (School of Chemistry and Chemical Engineering, Chongqing University, Chongqing, China PR)	122170
	Atmosphere Corrosion Assessment of Stainless Steels in Wenchang Coastal Area of Hainan Island	Cong Liu (Southwest Technology and Engineering Research Institute, Chongqing, China PR) / Lunwu Zhang (Southwest Technology and Engineering Research Institute, Chongqing, China PR) / Wei Liu (China South Industries Group Co., LTD., Chongqing, China PR) / Wanjun Yang (Southwest Technology and Engineering Research Institute, Chongqing, China PR) / Shiyan Zhang (Southwest Technology and Engineering Research Institute, Chongqing, China PR) / Bingfei Hu (Southwest Technology and Engineering Research Institute, Chongqing, China PR)	122210
CO2-Corrosion in CCS-Applications (TF)			
TF2.1	Corrosion behavior of carbon steel in presence of ferrous carbonate scales in supercritical CO2-water environments	Xiaodan Wang (State Key Laboratory of Multiphase Flow in Power Engineering, Xi'an Jiaotong University, Xi'an, China PR) / Yueshe Wang (State Key Laboratory of Multiphase Flow in Power Engineering, Xi'an Jiaotong University, Xi'an, China PR)	115132
TF2.2	Understanding H2 Gas Breakout in Carbon Capture and Storage Pipeline Gathering Networks	Thomas Jones (Pace Flow Assurance, London, United Kingdom) / Matthew Healey (Pace Flow Assurance, London, United Kingdom)	120762
TF2.3	The Influence of Ionic Strength on CO2 Corrosion of Mild Steel	Tatiana C. Almeida (NDC/COPPE/UFRJ, Rio de Janeiro, Brazil) / Rafael F. Garcia (NDC/COPPE/UFRJ , Rio de Janeiro, Brazil) / Merlin C. E. Bandeira (NDC/COPPE/UFRJ , Rio de Janeiro, Brazil) / Rogaciano M. Moreira (NDC/COPPE/UFRJ , Rio de Janeiro, Brazil) / Oscar R. Mattos (NDC/COPPE/UFRJ , Rio de Janeiro, Brazil)	121770

Posternumber	Paper title	Authors	Submission ID
Accelerated Atmospheric Corrosion Testing – Exchange Platform for Advanced Methods (WP6, WP8, WP14, WP17, WP22 & WP23)			
WP6-8-14-17-22-23.1	Comparison of the influence of water on degradation of poly (ethylene terephthalate) between indoor accelerated weathering and outdoor exposure tests	Azusa Ishii (NTT Device Innovation Center, NTT Corporation, Atsugi-shi, Kanagawa Pref., Japan) / Takashi Miwa (NTT Device Innovation Center, NTT Corporation, Atsugi-shi, Kanagawa Pref., Japan) / Hiroshi Koizumi (NTT Device Innovation Center, NTT Corporation, Atsugi-shi, Kanagawa Pref., Japan)	107537
Corrosion issues in CO2 environment (WP13 & TF)			
WP13-TF.1	Effect of carbon steel microstructure on FeCO3 precipitation kinetics during CO2 corrosion in the annulus of flexible pipes	Tatiane Oliveira Campos (Labcorr, UFRJ, Rio de Janeiro - RJ, Brazil) / Emanuel Seixas Nascimento Filho (Labcorr, UFRJ, Rio de Janeiro-RJ, Brazil) / José Antônio da Cunha Ponciano Gomes (Labcorr, UFRJ, Rio de Janeiro-RJ, Brazil)	106872
Corrosion protection by ZnAlMg coatings in atmospheric conditions (WP14, WP17 & TF)			
WP14-17-TF.1	A study on corrosion performance of Zn based alloy coated steel sheet under outdoor exposure testing	Kyung-Hwang LEE (Research Institute of Industrial Science & Technology, Pohang, Korea, South) / JiHoon YANG (Research Institute of Industrial Science & Technology, Pohang, Korea, South) / JaeIn JEONG (Research Institute of Industrial Science & Technology, Pohang , Korea, South) / Jeonghyeon YANG (Gyeongsang National University, Tongyeong, Korea, South) / InYoung PARK (Gyeongsang National University, Tongyeong, Korea, South)	109757
WP14-17-TF.2	IMPROVING CORROSION RESISTANCE OF STEEL WIRES WITH MAGNESIUM REINFORCED ZN-%5AL ALLOY COATING AND OBSERVATION OF MICROSTRUCTURE CHANGES	TESLIME AYMAN (R&D, ADANA, Turkey) / MUSA EMRE OZTURK (R&D, ADANA, Turkey)	114417

Posternumber	Paper title	Authors	Submission ID
WP14-17-TF.3	SVET and SIET study of galvanic corrosion of Al/MgZn2 in aqueous solutions at different pH	Alexander Immanyikwa Ikeuba (Institute of Metal Research, Chinese Academy of Sciences (IMR, CAS) Shenyang China PR) / Bo Zhang (Institute of Metal Research, Chinese Academy of Sciences (IMR, CAS) , Shenyang China PR) / Jianqiu Wang Wang (Institute of Metal Research, Chinese Academy of Sciences (IMR, CAS) , Shenyang China PR) / En-hou Han (Institute of Metal Research, Chinese Academy of Sciences (IMR, CAS) , Shenyang China PR) / Wei Ke (Institute of Metal Research, Chinese Academy of Sciences (IMR, CAS) , Shenyang China PR) / Peter Okafor (Department of Pure and applied chemistry, University of Calabar, Calabar, Nigeria)	122095

Polymers in organic coatings (WP14 & WP19)

Posternumber	Paper title	Authors	Submission ID
WP14-19.1	Effect of 2D MoS2/SDBS on anticorrosion performance of epoxy coating	Li Hongling (Institute of Oceanology, Chinese Academy of Science, Qingdao, China PR) / Xia Zhao (Institute of Oceanology, Chinese Academy of Science, Qingdao, China PR) / Baorong Hou (Institute of Oceanology, Chinese Academy of Science, Qingdao, China PR)	103587
WP14-19.2	Study on Correlation between Aging Failure of Acrylic Protective Coating and Atmospheric Environment in the South China Sea	jin gao (Institute for Advanced Materials and Technology, University of Science and Technology Beijing, Beijing, China PR) / zhen lv (Institute for Advanced Materials and Technology, University of Science and Technology Beijing, Beijing, China PR) / rui wang (Institute for Advanced Materials and Technology, University of Science and Technology Beijing, Beijing, China PR) / dawei zhang (Institute for Advanced Materials and Technology, University of Science and Technology Beijing, Beijing, China PR) / xiaogang li (Institute for Advanced Materials and Technology, University of Science and Technology Beijing, Beijing, China PR)	104312
WP14-19.3	CHARACTERIZATION OF THE PHENOMENON OF AGING PAINTING ANTI-CORROSION BASED ON EPOXY RESIN	Yazid Touazi (EMP, Adendorf , Kr Lüneburg)	117602

Durability of photovoltaic systems

Posternumber	Paper title	Authors	Submission ID
	Corrosion Behavior of Brystalline Silicon Solar Cells	XUETAO LUO (Xiamen University, XIAMEN, China PR) / LIUQING HUANG (XIAMEN UNIVERSITY, XIAMEN, China PR) / HUAPING XIONG (XIAMEN UNIVERSITY, XIAMEN, China PR)	101811

Exhibitors

Jotun

Sandvik

SHERWIN-WILLIAMS

Sika

PPG

POLSKIE TOWARZYSTWO CYNKOWNICZE

Polyguard

AMETEK Scientific Instruments

Armacell Engineered Systems

Automa s.r.l.

BAC France

cescor s.r.l.

ClampOn AS

Cormet Testing Systems

Cortec® Corporation Europe

EQUILABRIUM

Gamry Instruments

Hempel

IPS Elektroniklabor GmbH

OLYMPUS

PalmSens BV

Pine Research

Q-Lab
Corporation
Europe Ltd.

RICE GRTgas

SETARAM

Tikkurila

Effect of surface preparation on high temperature oxidation kinetics and oxide scale morphology

Wojciech NOWAK[1], **Bartek WIERZBA[1]**, **Jan SIENIAWSKI[1]**

[1] *Rzeszow University of Technology, Poland*

The materials used at high temperature, like in gas turbines or jet engines, need to fulfill a number of requirements, e.g. high creep strength and oxidation resistance at a wide range of operating temperatures, environments and loading condition as well as a suitable ductility at low temperature. Such properties obey Ni-base superalloys, due to their microstructure consisting of thermodynamically stable γ-Ni matrix with combination of strengthening γ'-Ni_3Al phase. However, during exposure of the alloys at high temperature, the oxidation process occurs and the material starts to form an oxide scale. The Ni-Cr-Al based alloys can be classified into the three groups of materials in term of formed oxide scales: NiO-forming, chromia forming and an alumina forming alloys. Formation of desired protective oxides like Al_2O_3 or Cr_2O_3 substantially increase the lifetime of the component exposed at high temperature. One of the way to provide a resistance against oxidation is applying of a protective coatings such as MCrAlY (where M is mainly Ni or Co) or β-NiAl which are an alumina forming materials are applied. However, coatings production is time consuming, results in additional component costs, and can negatively affect alloy mechanical properties, such as fatigue strength. Therefore, another, cheaper method to force material to form a protective oxide scale is proposed in the present study. Namely, a different surface preparation of the alloys, like grinding, polishing, sand blasting etc. on oxide scale formation during air exposure at high temperature will be presented.

Relationship Between Chemcial Structure and Scale Inhibition Efficiency of Polymerci Antiscalants

Wolfgang HATER[1], Duygu DISCI-ZAYED[1], Deborah LEANEY[1], Florian WOLF[1]

[1] *Kurita Europe GmbH, Germany*

Scale formation is wide-spread and inevitable problem in all kinds of water bearing systems, which is the cause scale-type and of multiple problems such as decreasing heat transfer, favoring corrosion or clogging of pipes, which might lead to reduced production or even plant shut down.

In order to mitigate scale formation the addition of an anti-scalant into water is state of the art. It is important to choose the proper anti-scalant according to the local conditions in order to solve the problem successfully and economically.

The efficacy of anti-scalants strongly depends on its chemical structure and scale-type. For polymeric anti-scalants the choice of monomers, the monomer ratio and the molecular weight strongly influence the properties of the anti-scalant.

Numerous studies by academia and industry have been published on the mechanisms of scaling and scale inhibition as well as on the efficiency of anti-scalants against all types of scale forming salts. However, systematic studies on structure/performance relations are hardly available.

The purpose of this contribution is to identify chemical structure-inhibition efficiency relationships of polymeric anti-scalants for various scale types encountered in water bearing systems.

In this track, the inhibition performance of selected homo-, co- and ter-polymeric anti-scalants, as a function of chemistry will be compared. Additionally, it will be analysed if the dependencies found are valid for single or multiple scale-types. The knowledge of the interdependencies enables the development of cost effective solutions to eliminate scale problems in water bearing industries.

The Influence of Charge Density of Anionic Macromolecules in Magnesium Silicate Inhibition: Relevance to Geothermal Systems

ARGYRO SPINTHAKI[1], Konstantinos DEMADIS[1], Duygu DISCI-ZAYED[2], Wolfgang HATER[2]

[1] *University of Crete, Greece*
[2] *Kurita Europe GmbH, Germany*

The Influence of Charge Density of Anionic Macromolecules in Magnesium Silicate Inhibition: Relevance to Geothermal Systems

Argyro Spinthaki,[#] Duygu Disci-Zayed,[§] Wolfgang Hater[§] and Konstantinos D. Demadis[#]

[#]Crystal Engineering, Growth and Design Laboratory, Department of Chemistry, University of Crete, Voutes Campus, Heraklion, Crete, GR-71003, Greece

Kurita Europe GmbH, Giulinistrasse 2, 67065 Ludwigshafen, Germany

Magnesium silicates constitute a family of geological formations both widely studied and well characterized. However, although this term is commonly used by water treatment professionals to indicate any oxidic deposit that contains both Si and Mg, the true identity of such deposits remains elusive. In reality, these water-formed deposits are actually amorphous silica with Mg^{2+} ions entrapped in its colloidal matrix and they bear little resemblance to their geological counterparts. It was reported that Mg^{2+} ions actually play a catalytic role in silica polycondensation chemistry [1].

The formation, precipitation, and deposition of the so-called "magnesium silicate" in geothermal waters have been subjects of intense interest. Such scaling poses a severe threat to the smooth process of industrial systems. The aim of our approach is the systematic study of the influence of chemical additives in the presence of magnesium ions and their varying results based on their chemistry.

Our studies are based on characterization of the "magnesium silicate" scaling and inhibition experiments carried out in supersaturated solutions of different silicate and magnesium concentrations at pH values from 7.0 up to 10.0. Anionic additives were found to act as "magnesium silicate" controlling agents, most likely by complexing the magnesium cations, and thus preventing "magnesium silicate" formation. Based on a plethora of experimental data, a number of useful functional insights have been generated, which add to building a more complete and comprehensive picture of the mechanism of "magnesium silicate" formation and control.

[1] K.D. Demadis, A. Ketsetzi, E. Sarigiannidou, Ind. Eng. Chem. Res. 2012, 51, 9032−9040.

Silica Scale Inhibitors: Current State-of-the-Art

Kostas DEMADIS[1], Giannis ARISTODIMOU[1], Argyro SPINTHAKI[1]

[1] *University of Crete, Greece*

Raw waters used as feedwaters for industrial cooling water systems contain a multitude of alkaline earth metal cations such as calcium, magnesium, barium, etc. These, particularly calcium, cause enormous problems when combined with polyanions, such as carbonate, phosphate or sulfate, while the solubility products of the corresponding sparingly soluble salts are exceeded. Major problems that ensue are precipitation and deposition of these insoluble mineral salts onto critical equipment surfaces (eg. membranes, heat exchanger tubing, tower fill, etc.). Calcium carbonate and calcium phosphates are the most frequently encountered deposits. Other deposits, although less common, are equally troublesome. These include silica and magnesium, aluminum or ferric(ous) silicate. Accumulation of the above deposits greatly diminishes effective heat transfer rates. These phenomena can cause catastrophic operational failures in water systems that may lead to costly downtime. Prevention of scale formation is greatly preferred by industrial water users to the more costly, laborious (and potentially hazardous) chemical and mechanical cleaning of the adhered scale in the aftermath of a scaling event.

This paper will focus on colloidal silica fouling and its control either by inhibition or by dissolution. The use of several designed treatment additives will be described, with emphasis on "green" scale inhibitors and cleaners. Silica scale inhibitors described can maintain up to ~ 400 ppm soluble silica in solution. Attention will be paid to the designed chemical approaches that should be used to combat silica scale. A detailed analysis will be given on the entire chemical treatment process, starting from proper selection of the treatment program, to proper application, to effective monitoring and maintenance of system performance. Also the general scope of silica fouling fundamentals will be presented.

Phosphorous- and heavy-metal-free corrosion inhibition for open cooling systems based of filming amines

Juergen MATHEIS[1], Wolfgang HATER[1], Florian WOLF[1]

[1] *Kurita Europe GmbH, Germany*

At present, corrosion inhibitors for mild steel in open cooling systems are mostly based on zinc and phosphorous compounds or combinations of both. Molybdates are rarely used for cost reasons. Since the blow-down of open cooling systems is normally discharged into rivers or lakes, environmental concerns are drivers to look for suitable alternatives.

In boilers and also closed cooling systems, corrosion inhibitors based on filming amines are getting increasingly established, raising the question if these could be an option also for open cooling systems, both with regards to performance and to the environment. A special focus was put on the efficiency in soft water (Ca^{2+} = 1 mmol/L = 100 ppm $CaCO_3$), where concepts based on phosphates and phosphonates already lose performance due to the low calcium content.

Corrosion inhibition was studied by means of electrochemical methods, laboratory circulation devices and pilot cooling systems. In pilot plant trials, corrosion rates < 0.05 mm/a on mild steel and the absence of pitting were achieved, test methods were coupons (weight loss) and the linear polarization resistance (LPR). While dosing rates were between 20 and 30 mg/L as active ingredient in the recirculating water, the concentration found by analytical means was less due to adsorption and degradation processes. Results could be further improved when combining the filming amines with selected co-inhibitors.

Current trials seek to identify filming amines with the best environmental profile possible. Further work deals with means to mitigate bacterial degradation and overly foaming during application, also with the influence of typical cooling water parameters, e.g. chlorides, pH, temperature.

Protective Coatings for High-Temperature Steam Oxidation in Coal-Fired Power Plants

Tomasz DUDZIAK[1], Eugene MEDVEDOVSKI[2]

[1] *Foundry Research Institute, Kraków, Poland, tomasz.dudziak@iod.krakow.pl*
[2]*Endurance Technologies Inc., Calgary, Canada, emedvedovski@shaw.ca; eugenem@endurancetechnologies.com*

Abstract
The corrosion resistance of protective coatings on stainless steels against high-temperature steam oxidation has been studied. The coatings are made through the thermal diffusion process and consist of aluminides with a 2-3 layer structure or with more complex architectures when the aluminide layers are covered by thin oxide-based (e.g. Al_2O_3, SnO_2, ZrO_2) or BN layers. The oxidation behaviour of these coatings compared to bare materials (e.g. 310SS and 800H) was studied at 800 °C for 3000 hours. In this work, along with determination of mass change of the samples vs. time (kinetic of oxidation), extensive structural analysis and surface hardness determination have been carried out. The developed multi-layered coating architectures demonstrated promising high-temperature steam oxidation performance with no breakaway degradation issues. Their performance is defined by the inertness of the aluminium intermetallides with a total thickness of protective coatings of greater than 100 µm, which are formed through the thermal diffusion process, with high diffusion-related bonding between the layers and with the substrate, and by the formation of the aluminium oxide films occurred due to the surface oxidation. The additional layer from the selected oxides or BN applied on the top of the aluminized structure can provide additional corrosion protection; however, it is important to maintain a high level of bonding between this layer and the intermetallide structure. The thermal diffusion technology and obtained coating systems are applicable for the protection of products with different sizes and shapes, including long tubing required in industry.

Keywords: steam oxidation, steel, high temperature corrosion, protective coatings

Introduction

Increase of efficiency and reduction of CO_2 emission are key factors of the service of coal-fired power plants. These can be attained by increase of operation temperatures and pressures in the water steam systems towards the "ultra- and advanced ultra-supercritical" conditions, where temperatures of steam greater than 720-750 °C are employed. These operation conditions require the products, e.g. long tubing, which should withstand the required corrosion conditions. Steel materials used in operation are generally adequate to the steam pressure and temperature. Thus, low grade steels like T/P22 (2.5 wt.-% Cr), T/P91 (9 wt.-% Cr), are used with limited functionality and temperature of below 600 °C because of quick oxidation and degradation, which are related only to the Sub-critical conditions. More advanced operation conditions (temperatures and pressures) require more advanced materials with high corrosion resistance. Steel materials with Cr contents of 2-10% cannot withstand the Ultra-Supercritical (USC) and Advanced Ultra-Supercritical (A-USC) service conditions due to their quick oxidation and degradation as is described in literature [1, 2]. Currently, particular attention is focused on the austenitic steels with a high Cr content (>20 wt.-%), such as 310 stainless steel (310SS) or more advanced with higher Ni contents, like Incoloy, where adequate mechanical properties are combined with high temperature corrosion resistance owing to the formation of continuous, thin, adherent, thermally grown oxide Cr_2O_3 and/or spinel-type scale, depending on metal composition [3, 4]. Nevertheless, under long term exposure even 310SS and other Cr- and Ni-rich steels and alloys experience degradation due to breakaway oxidation [5, 6]. For these processes, tubing and components should be made of highly expensive Ni- and Cr-rich alloys (which have limited performance), or they should be protected with high-temperature corrosion resistant coatings.

Although some coating materials and technologies (e.g. cold and thermal spray, PVD, electrophoretic deposition) may be considered for high-temperature oxidation protection, many of them are hardly applicable for complex shape and large-size products, e.g. for long tubing (up to 10 m length) when inner or inner and outer surfaces need to be protected. In this regard, aluminized coating on steels can be of a highly potential since they can withstand high-temperature corrosive environments and can be applicable for large and complex shape industrial products. These coatings consist of two or more layers of aluminides (iron aluminides, iron chrome aluminides and some others depending on the metallic substrate and process parameters). They are produced through the thermal diffusion technology (thermo-chemical processing or pack cementation) according to the CVD principles and are suitable for variety of steels and Ni-based alloys.

The evaluation of the materials for high-temperature oxidation, in general, requires substantial time, especially for the alloys with high contents of Ni, Cr and some other metallic ingredients and for the coated steels and alloys. Thus, the long-time testing for 2000 - 3000 hours or longer, even up to 10000 hours will provide reliable results and appropriate understanding of the material performance. In the present work, a few different coating structures (architectures) based on the intermetallide phases in the Fe-Al, Cr-Al, Ni-Al systems deposited on 310SS and 800H Incoloy through the thermal diffusion technology have been prepared and tested in the high-temperature steam oxidation condition (at 800 °C for the duration up to 3,000 hrs) with consequent surface and structure evaluation.

Materials and Methods

Starting materials and Coating deposition process

Stainless steel 310 grade (310SS) and Incoloy 800H produced by Sandvik Materials Technology and ThyssenKrupp VDM and supplied by Rolled Alloys (Canada) have been selected for the study. **Table 1** shows chemical compositions of these starting materials according to the provided Mill Test Report (MTR) certificates. The materials have been treated using a 600-grit SiC paper and ultrasonically cleaned with acetone for 15 min.

***Table 1** Chemical composition of the uncoated 310SS and 800H (wt.-%)*

	Fe	Ni	Cr	Mo	Si	Mn	Nb	Al	C	P	S	Ti
310SS	Bal.	20.11	24.51	0.42	1.5	1.25	0.01	0.005	0.043	0.022	0.0009	-
800H	46.6	30.3	20.6	-	0.4	0.7	-	0.44	0.09	-	<0.002	0.51

For the coating formation, the samples were processed according to the proprietary aluminizing technology (EndurAlon™) of Endurance Technologies Inc. (ETI). This technology is one of the options of thermal diffusion processing or pack cementation. The steel samples cleaned from a grease and a scale, as described above, have been placed in a special retort contained a special aluminized mix. This mix contained an Al-rich donor, a halide salt used as an activator, and inert filler; all materials were in the powder form. The mix formulation was developed by ETI, and particle size of the starting powders has been specially selected providing easy flow and compaction. The sealed retort with the aluminized mix and the samples has been heated in the temperature range of 850 – 1000 °C for a certain holding time period in an electric kiln. At heat treatment, the Al rich gases are formed due to high-temperature reactions; the forming atoms of Al deposit onto the metallic substrate surface with consequent diffusion to the steel structure. Due to the inward diffusion of Al and outward diffusion of Fe, Cr, Ni, the aluminides are formed with subsequent consolidation of the coating structure. Upon the process, three aluminide protective layers with different contents of Al (related to the formation of certain aluminides) are formed simultaneously on the metallic surface. After cooling of the kiln, the samples were removed from the processed powder and cleaned using a brush and acetone.

In order to prepare the coatings with more "complex" architectures, an additional protective layer was applied onto the aluminized samples. To obtain that, the aluminized samples were immersed into the ceramic suspensions with nano-particles of selected concentrations; the top layer deposition was conducted under pressure. These suspensions are based on ceramic materials with high oxidation resistance, e.g. based on the oxides of metals of III-IV group of the Periodical Table and BN, i.e. based on the materials with rather high degrees of covalent bonds and thermodynamic potentials. Thus, for the top coating, the ceramic slurries contained Al_2O_3, SnO_2, ZrO_2 (3 mol.-% Y_2O_3) and BN have been utilized; the solid contents varied from ~0.5 wt.-% to 30 wt.-%. The samples with applied coatings have been air dried and then consolidated in an electric kiln at temperatures ranging 500 - 700 °C. The types of the coatings are shown in **Table 2**. The approximate dimensions of the samples with coatings for oxidation corrosion evaluation were (10-17)x(10-12) x6.3 mm. The dimensions of the uncoated samples, which were cut from small tube sections, were smaller.

Table 2 *Types of the Coatings Tested*

Sample code		Description
3A	8A	Aluminized steel
3B	8B	Aluminized steel with Al_2O_3-based top layer
3C	8C	Aluminized steel with SnO_2 top layer
3D	8D	Aluminized steel with ZrO_2-based top layer
3E	8E	Aluminized steel with BN top layer

*Number 3 denotes 310SS and number 8 denotes 800H

Experimental Procedure

The schematic of high-temperature steam oxidation test rig is presented in **Figure 1**. In this type of steam oxidation test set-up [7], 100% pure steam is generated by pumping water from a reservoir placed underneath the furnace. Water steam passes over the test samples with a flow rate of 2.833 ml/min (1 rotation per minute (rpm) and flows into a condenser before the water returns to the reservoir. The water used in the reservoir was ~~is~~ de-ionised twice. The whole system was sealed using stainless steel flanges from both ends, prior to the test. The test chamber was purged using an oxygen-free nitrogen flow. This purge continued through the water reservoir throughout the samples for 2 hrs exposure period to minimize the level of oxygen in the system. Prior to the exposure, the furnace has been calibrated to ensure the placement of the samples in the furnace at a test temperature with accuracy ± 5 °C. In this study, the test has been conducted at 800°C for totally 3000 hours. The samples have been inserted into the hot zone of the furnace using a calibrated stick with the measured size, placed them on the Al_2O_3 plate within the hot zone. During the oxidation testing, the samples were withdrawn from the hot furnace after certain periods of time (e.g. 200, 400, 750, 1200, 1700, 2200, 3000 hrs), checked them for the mass change, and then placed again into the hot zone of the furnace for continuation of the oxidation.

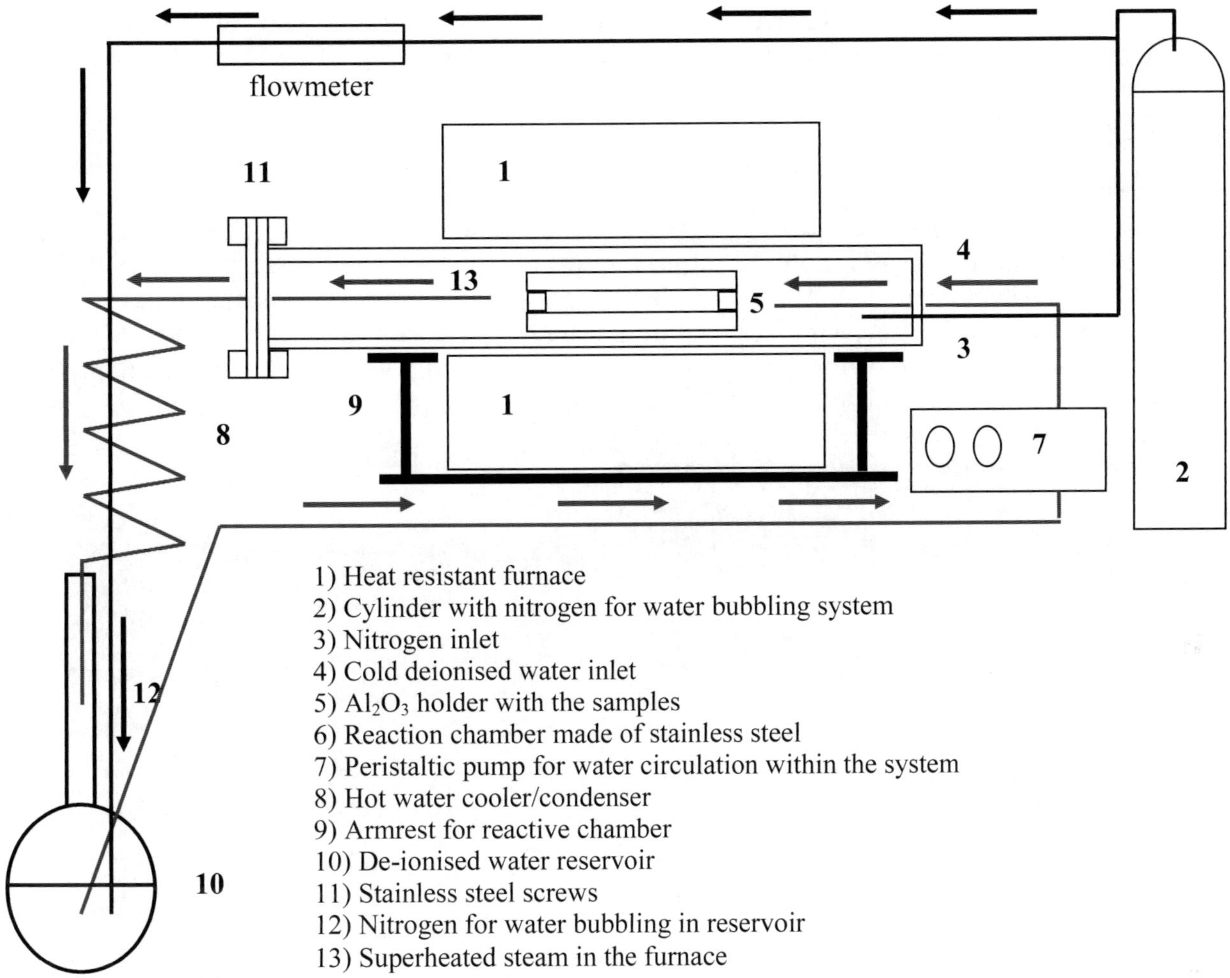

***Figure 1** Steam rig used in the high-temperature oxidation study*

Prior and during the exposure, the samples were accurately weighed using a digital balance with a resolution of ± 0.01 mg for masses m = < 80 g. The digital balance (Sartorius CPA225D) was calibrated frequently using its internal calibration function and periodically with test weights. Prior to exposure and during micro examinations samples were handled with a care using nitrile gloves in order to avoid surface contamination.

<u>Materials examinations</u>

The samples prior and after exposure have been visually examined from the surfaces using DSLR Canon EOS 70D camera coupled with Canon MP-E 65mm f/2,8 Macro lens. Scanning Electron Microscope (SEM) Hitachi TM3000 coupled with Bruker Energy X-Ray Dispersive Spectrometry (EDS) have been utilized for the surface and chemical composition detailed evaluation. The SEM examination for the surface morphology (before and after oxidation testing) was conducted in the Backscatter Electron (BSE) mode under 15 kV accelerated voltage. The examination of the cross-sections has been performed with light optical microscope MEIJI Techno 1M7200 to observe possible changes of the coating structures. For the samples preparation for the cross-section microscopy examination, precision cutting-polishing multiple-step procedure established at ETI has been employed using the Buehler equipment. The special etching composition has been selected to reveal the structural features in the cross-sections. Micro-hardness determination of the aluminized coatings have been conducted according to ASTM E384 with the Clark CM400AT tester applying the Knoop diamond indenter to the polished cross-section at the load of 100 g (HK0.1).

Results and Discussion

Coating Appearance

The appearance of the samples, after the oxidation testing, is shown in **Figure 2A** (310SS substrate) and **Figure 2B** (800H substrate), respectively. The samples showed no spallation, cracks, delamination or any other visible defects. Small spots on some samples can be observed, and these spots with local colour deviations on the surface may be related to uneven thickness of the top coatings (which were applied by a dipping technique) and imperfect surface of the aluminized samples, on which these top layers were applied. Some discoloration has only sacrificial character.

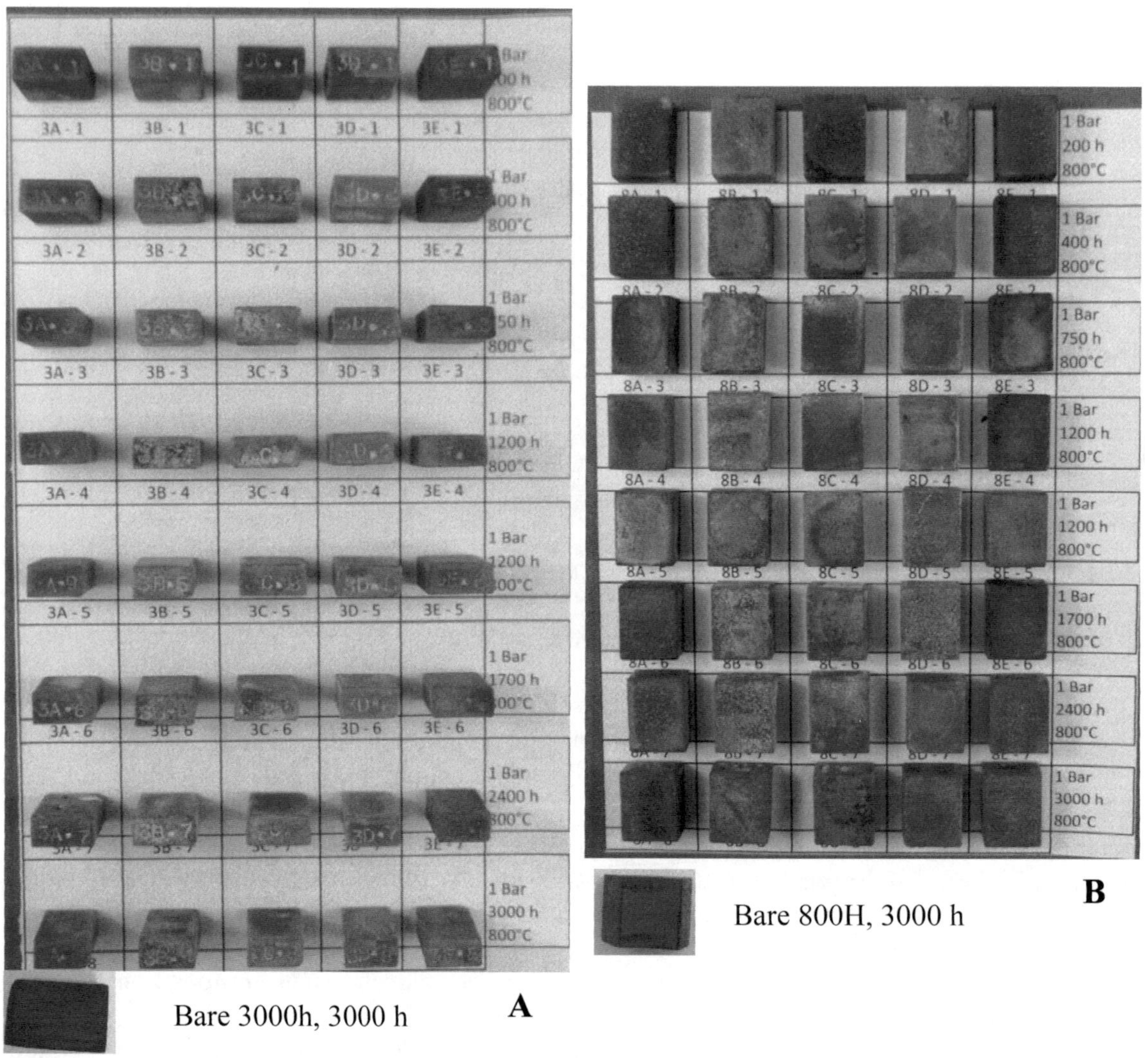

***Figure 2** Appearance of the coated samples after steam oxidation tests A) 310SS and B) 800H substrate*

Oxidation kinetics

The graphs in **Figure 3** show "kinetic data" of the selected materials exposed for up to 3000 hours at 800 °C in steam atmosphere. The mass change (gain) per the surface area has been calculated. The plotted curves (mass change vs. exposure time) demonstrate a "parabolic" behaviour. The mass gain occurs rather intensively during the first 800-1000 hrs, approaching the "plateau", and then the mass of the samples practically did not change. Because of that, the further high-temperature exposure was not very reasonable. The mass gain for the samples with the aluminized coatings is greater compared to the uncoated materials. The mass gain is defined by the surface oxidation of the aluminized coating, e.g. by the oxidation of aluminides and the related formation of aluminium oxide. Depending on the thin top coating layers (oxides or BN), the oxidation and related mass gain are for the samples are slightly different. Thus, the presence of the top layer above the aluminized coating reduced the mass gain (the samples 3A and 8A had higher mass gain than others). Comparing the samples with the top layer, it is difficult to make a clear conclusion, which top layer results in less or higher mass gain.

310SS 800H

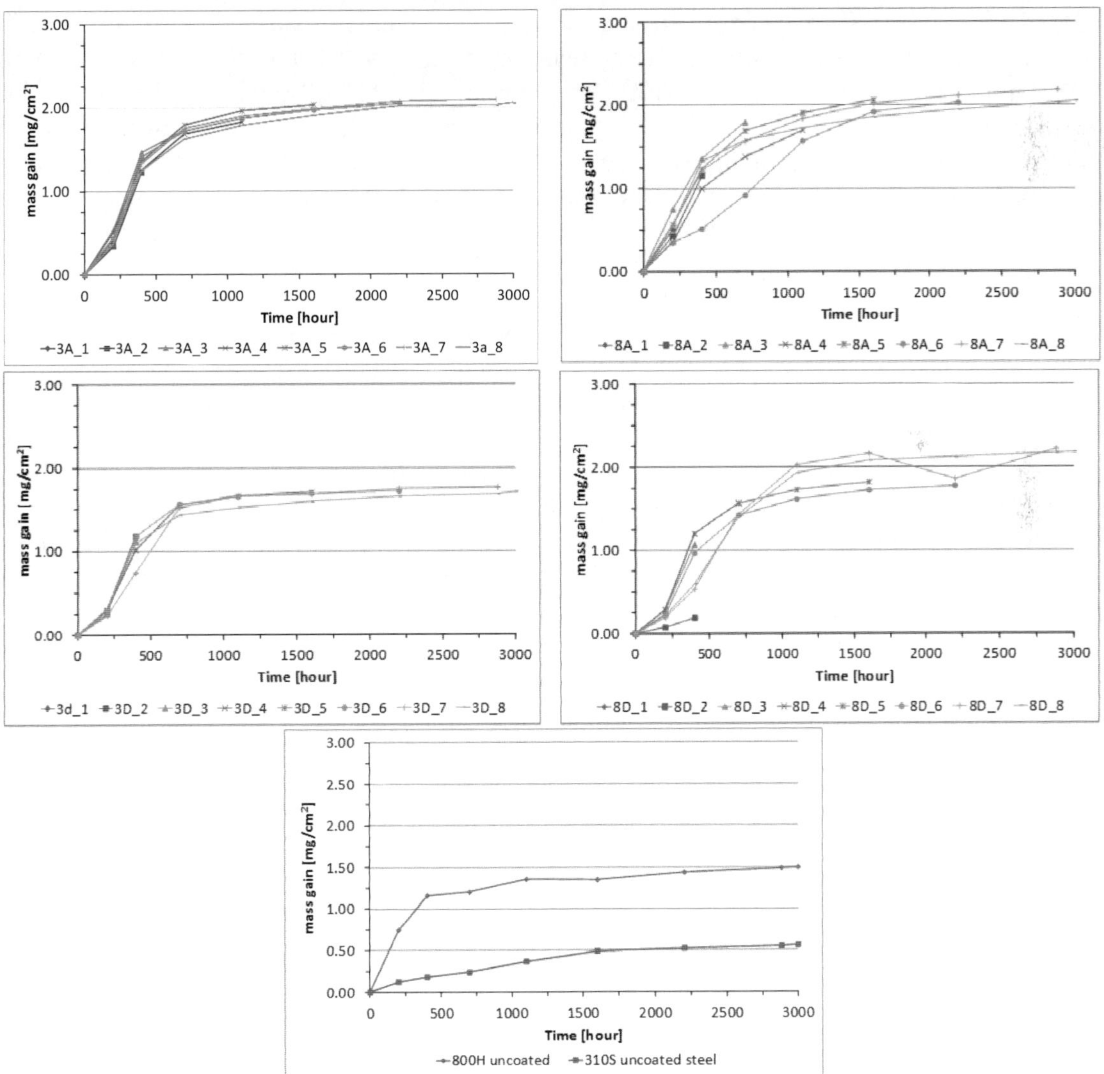

Figure 3 *Kinetic data of the samples exposed in steam atmosphere for up to 3000 hours at 800 °C*

A higher mass gain was detected for the uncoated 800H Incoloy than for the uncoated 310S steel, that may be related to lower Cr content in 800H compared to 310SS, and higher Ni concentration in 800H than in 310SS substrate.

The mass change and its behaviour during the exposure time cannot be a "straightforward" direction of the materials' performance (oxidation protection) since these values do not indicate the surface and structural changes, which may occur at the oxidation. In particular, in the case of bare steels, the formed oxidation scale may partially detach from the surface resulting in the mass loss. In this regard, the structural examination of the coatings and bare materials should be more important for the consideration of the materials' integrity.

Microscopic surface and structural examination

The surface morphology studies with SEM in the BSE mode conducted for the samples before and after steam oxidation test, showed that the bare materials (310SS and 800H) experienced the formation of scale and some spots, pinholes and small micro-cracks on the surface. Although 800H had significantly fewer spots, it experienced more visible delamination and detachment than 310SS (will be illustrated further). The change of surface morphology of the samples with aluminized coatings after steam oxidation is related, first of all, with the oxidation of the aluminides, i.e. with formation of the Al_2O_3-based surface "skin" and consolidation of the this "skin". This is confirmed by the data of the EDS analysis (the contents of O became significantly higher). As well known, Al_2O_3 has a high chemical inertness and stability at elevated temperatures, and its formation can be considered as a positive factor. The samples with only aluminized coating (3A and 8A) and with aluminized coating and a very thin BN top layer (3E and 8E) have more even surfaces, while the aluminized samples with oxide top layers (3B and 8B, 3C and 8C) have less homogeneity with an appearance of surface micro-cracks. Some deviations from even surface (e.g. small spots) of the aluminized samples and the samples with top BN layer are probably related to uneven morphology and thickness of the formed Al-rich layer and uneven pore distribution on this surface. For the samples with the top oxide layers applied (Al_2O_3, SnO_2 and ZrO_2), which thicknesses vary from below than 1 μm to several μm, the presence of the "egg-shell" surface micro-cracks observed is probably dealt with the adhesion issue occurred at thermal treatment after the top oxide layer deposition; e.g. with a coefficient of thermal expansion mismatch of the oxide layers and the aluminized steel. In addition, the tensile stresses occurred in the top coating layer during its consolidation at the heat treatment. However, these surface micro-cracks are well visible only at high magnifications (500-1000x or greater); their width are rather small, below 1 μm, in particular, ranging from 70 to 300 nm (depending on the oxide layer composition), and their depth is also below 1 μm. The top oxide coatings have dense structures without visible pits and pores. It can be noted that the top ZrO_2 coating remained intact after long-time high-temperature oxidation, while other oxide top coatings (Al_2O_3 and SnO_2) were partially detached that is related to worse bonding of these thing layers to the aluminized coating. The surface oxidation occurring at the corrosion testing and long exposure promote the healing of the micro-cracks. To illustrate the mentioned, the SEM images of the selected samples, e.g. 3A (aluminized), 3D (aluminized with a top ZrO_2-based layer) and 3E (aluminized with a top BN layer), as well as the uncoated 310SS, before and after high-temperature oxidation for 3000 hrs are demonstrated in **Figure 4**.

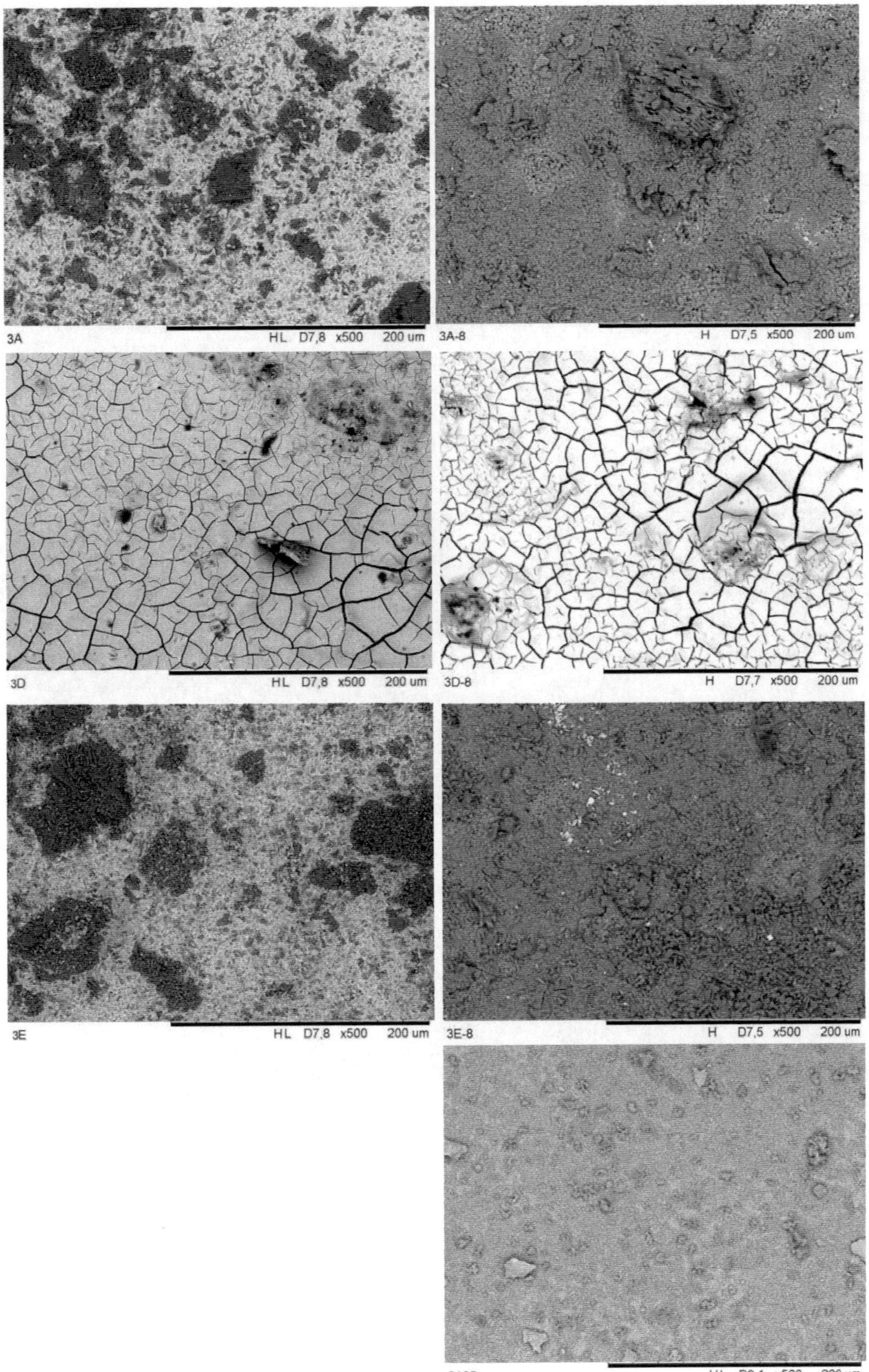

***Figure 4** The surface morphologies of the selected samples prior (on the left) and after (on the right) 3000hrs oxidation test: 3A, 3D, 3E and 310SS (uncoated)*

The cross-section microstructure images for the aluminized samples before and after 3000 hours exposed at 800 ^{o}C are shown in **Figure 5**.

310SS -A (Aluminized steel) before test

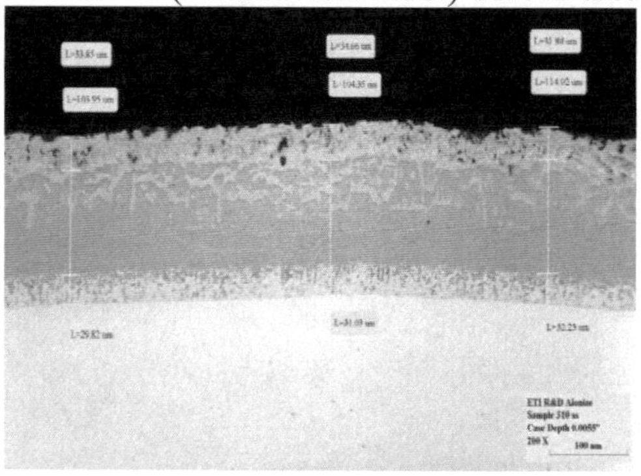

800H (Aluminized steel) before test

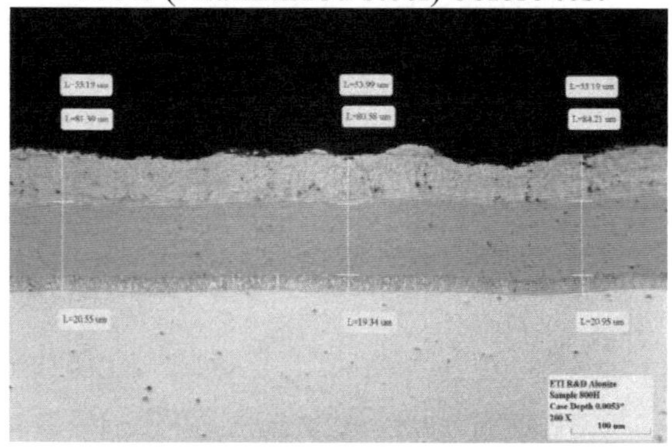

310SS - A (Aluminized steel)

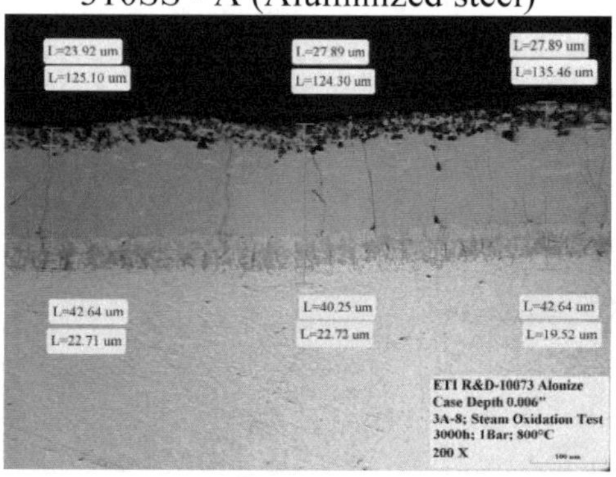

800H (Aluminized steel)

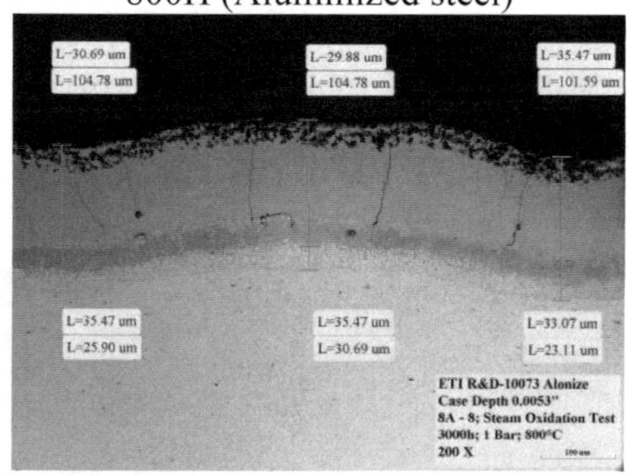

310S-B (Aluminized steel with Al_2O_3-based top layer)

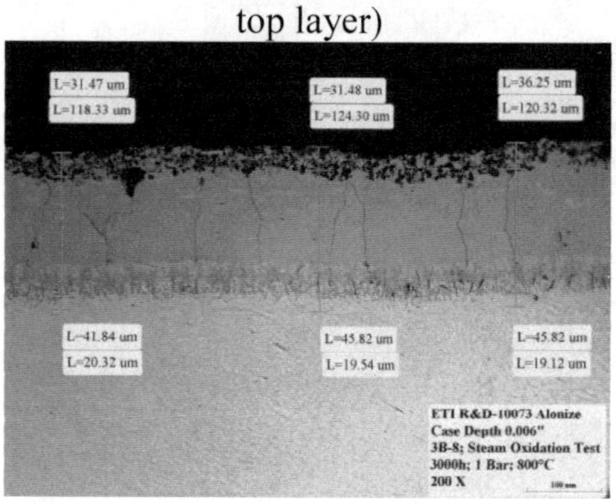

800H-B (Aluminized steel with Al_2O_3-based top layer

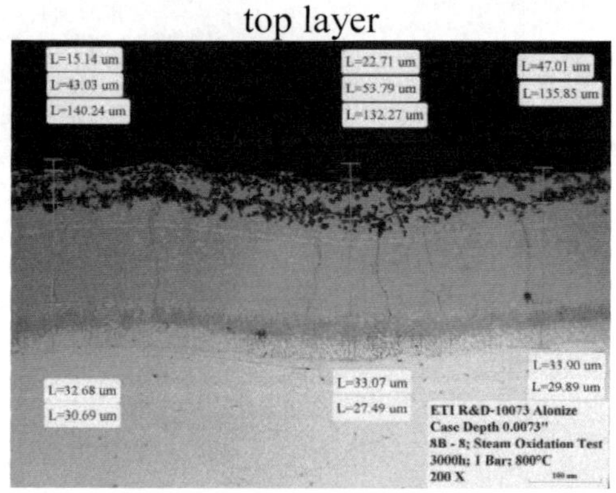

310S-B (Aluminized steel with Al_2O_3-based top layer)

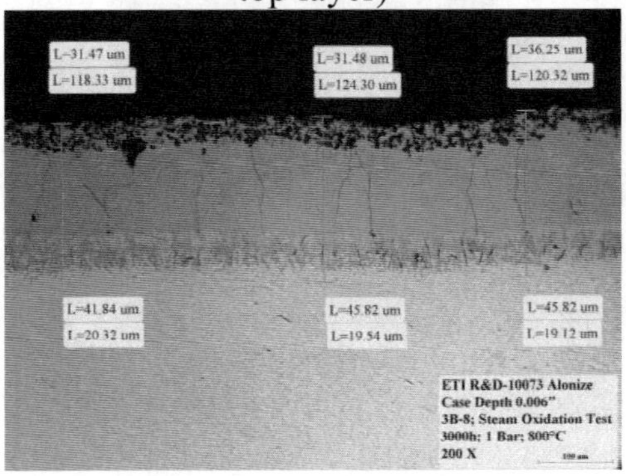

800H-B (Aluminized steel with Al_2O_3-based top layer

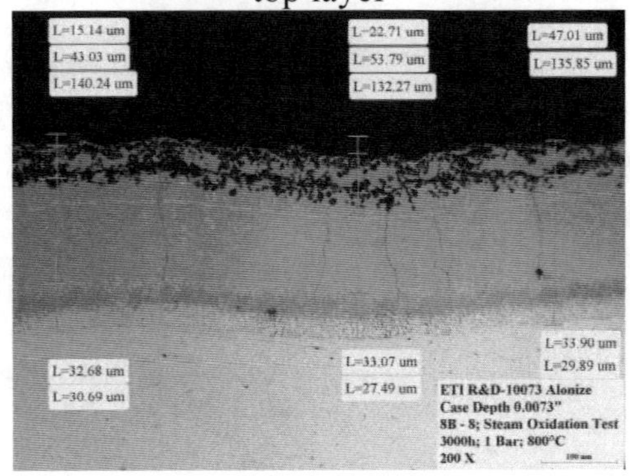

310S-C (Aluminized steel with SnO_2 top layer)

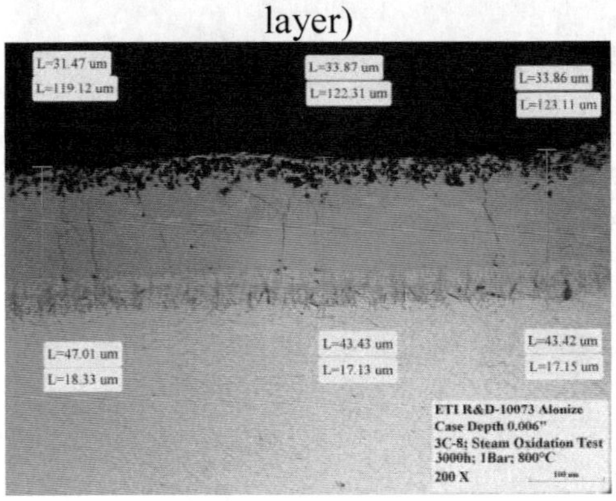

800H- C(Aluminized steel with SnO_2 top layer)

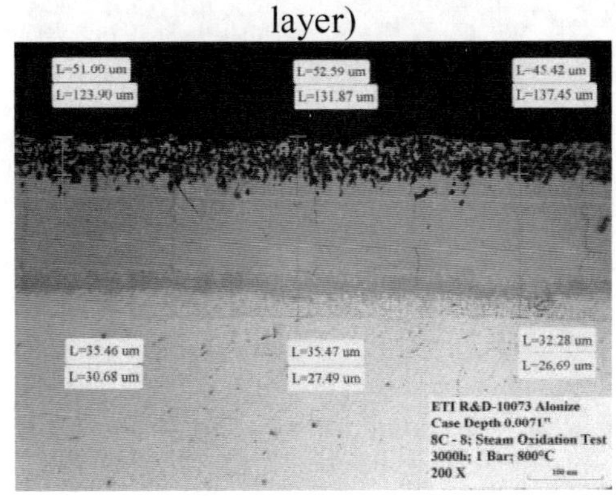

310S-D (Aluminized steel with ZrO_2-based top layer)

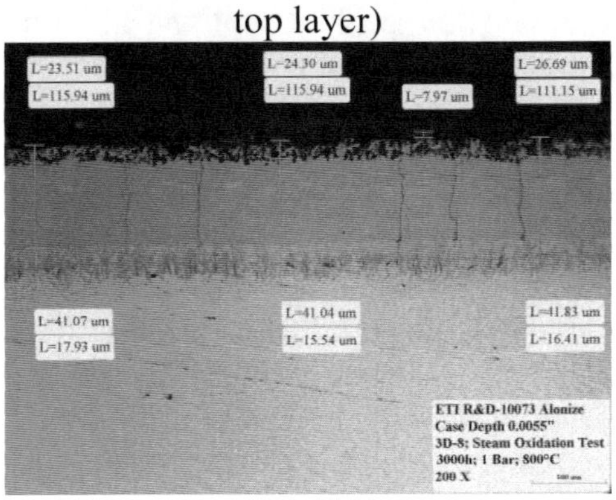

800H-D (Aluminized steel with ZrO_2-based top layer)

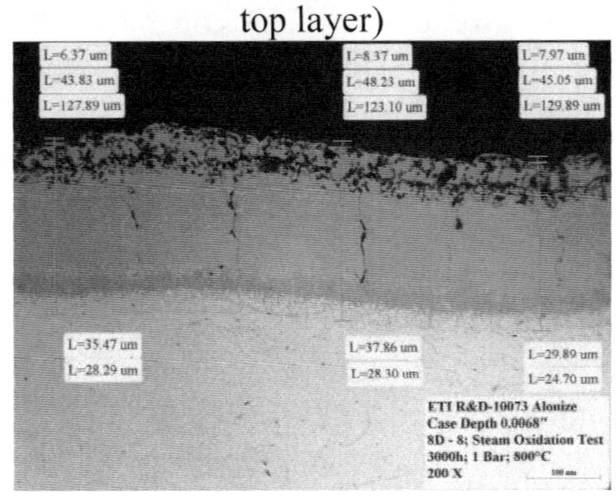

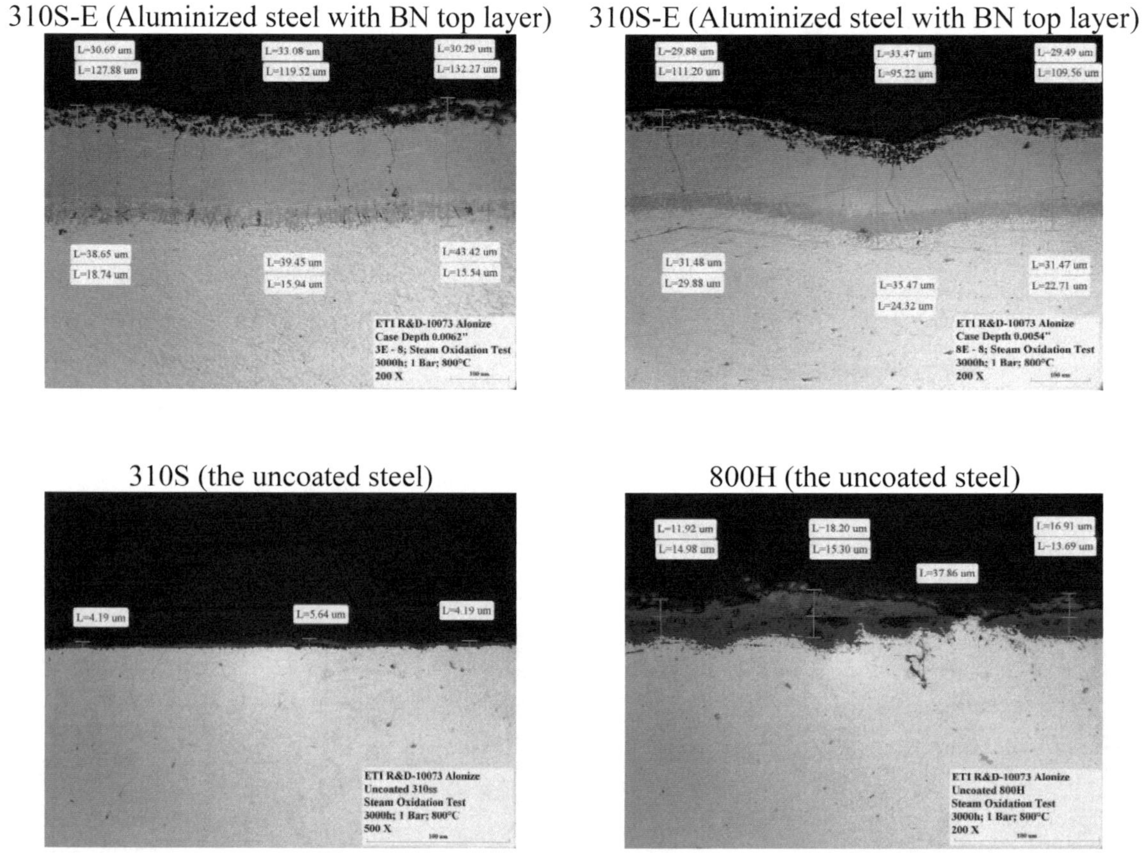

Figure 5 *Microstructure (cross – sections) of the exposed materials before (first row) and after oxidation testing (other rows)*

The observed cross-sections of the samples with the aluminized coatings revealed only insignificant changes. They can be dealt, first of all, with the oxidation of the aluminides leading to the formation of the chemically inert layer of Al_2O_3 phase and other oxides and its consolidation. No visible coating destruction, delamination or pitting and, moreover, the breakaway issue were observed. The appeared porosity in the upper zone of the coating is related to the oxidation and not enough consolidation in this zone. However, the thickness of the zone contained porosity is significantly less than the thickness of well-consolidated coating. The hairline micro-cracks observed on some coatings, especially after long exposure, are related to the thermal shock occurred when the samples were withdrawn from the furnace and then placed back several times. Anyway, the observed hairline cracks stopped at the transition zone for all the coated 310SS and 800H samples that indicates to satisfactory integrity of the coatings structures to the selected extreme testing conditions. It should be outlined that this testing conditions distinguish from the industrial steam oxidation service conditions when final temperatures and especially heating ramps are lower. In the industrial conditions where service temperatures are slightly lower, i.e. 600 – 700 °C, continuation of diffusion processes and related changes in the coating microstructures would be significantly minimized and thermal stresses would be also less. The diffusion-related bonding between the coating layers and with the substrate materials promotes the integrity of the aluminized coating structure at high temperature oxidation.

It is interesting to note that the total case depth of the aluminized coatings increased after the high temperature oxidation exposure. Furthermore, the Al-rich layer (with the Al contents of ~44-50 wt.-%) of the aluminized coatings practically disappeared for the aluminized 310SS samples, while this Al-rich layer became significantly smaller for the aluminized 800H samples. However, the main aluminized zone (with the Al contents of ~30-38 wt.-%), as well as the thickness of the transition zone (with the Al contents of ~3-7 wt.-%), became greater for all samples. Even the transition layer became with 2 sub-layers. Thus, after the oxidation testing, the main aluminized layer, which, in fact, provides corrosion protection, grew up to a thickness of 140-150 μm in the case of the 310SS substrate. In the case of aluminized 800H samples, a thickness of the main aluminized layer became 105-120 μm, and a thickness of the total protective layer (main and Al-rich layers) also became around 130-150 μm. This growth can be explained by the continuation of the diffusion processes at the high testing temperature (which is close to the temperature used for the aluminizing process) during the long-time exposure.

Despite the continuation of the diffusion processes (inward Al diffusion and outward diffusion of Fe, Ni, Cr), no Kirkendall porosity and micro-gaps between the layers of the aluminized coating occurred during the high-temperature oxidation test that should be outlined as a positive point. Some growth of the case depth during the high temperature exposure may affect partial destruction and flaking of the top thin oxide layers adherent to the coatings 3B, 3C, 3D and 8C and 8D. Comparing the samples with two different substrates, the aluminized coatings on both substrates demonstrated similar adhesion.

Determination of the micro-hardness (HK0.1) of the aluminide layers for the samples before and after oxidation testing revealed the high integrity of the coatings. The HK0.1 values remained on the same level after the testing. Thus, the HK0.1 values of the main aluminized layer determined for the samples before and after oxidation testing, regardless the substrate material, remained in the range of 600 - 650 kgf/mm^2 (only when the indenter was applied either close to the pores-contained upper zone or to the transition zone, the HK0.1 values dropped to 520-570). The hardness of the transition layer also remained in the range of 380-450 kgf/mm^2. The substrates experienced insufficient hardness reduction (from 250-270 to 200 - 250 kgf/mm^2 for 310SS and from 220 - 240 to 170 - 200 kgf/mm^2 for 800H). The micro- hardness of the scales formed on the bare materials could not be determined accurately, even at lower indentation loads, due to general "softness" of the scales and poor bonding to the substrates.

As opposed to the coated samples, bare 310SS and 800H experienced different behaviour. The 310S uncoated material showed the formation of a thin oxide scale composed mostly of $MnCr_2O_4$ phase and Cr_2O_3; on the other hand, the 800H steel showed the formation of much thicker oxide scale (25-35 μm or greater), with lowered adhesion (see **Figure 5**). The oxide scales contained large quantities of Fe, Cr and Ni. The 310SS sample exhibited uneven Cr contents on the surface (according to the EDS analysis, from 9 to 47 wt.-% depending on different areas of the surface) and some nodule formation, despite a high content of Cr in the metal matrix, that corresponds to the breakaway oxidation phenomena [8, 9] Significantly elevated contents of Mn ranging from 8 to 24 wt.-%, which also varied in different areas, can be related to the outward diffusion of Mn and to formation of spinel phases, like $MnCr_2O_4$ (the content of Mn in the steel is only 1.25% according to MTR). Thus, the Cr_2O_3 inner sub-layer and the $MnCr_2O_4$ outer sub-layer may "compose" the scale [9], but these layers are unevenly distributed that would affect the nodule and pitting formation. It should be noted, that due to rather poor adhesion of the formed oxidation scales and their "softness", there was a probability that these scales were partially detached and removed during the cutting and polishing at the cross-section samples preparation despite precision cutting saw and multi-stage polishing procedure applied. In this regard, the values of scale thicknesses can be considered as approximate.

As mentioned above, the aluminized samples demonstrated their adequate corrosion resistance against the steam condition at 800 ^{o}C. Comparing the formation of a thin alumina layer for the aluminized coatings during high temperature oxidation with a chromia scale formation occurring at this process for uncoated Cr-rich stainless steels, the former can be considered as preferable. It may be related to significantly slower formation of the aluminium oxide scale, according to a number of studies conducted with stainless steels with sufficient contents of Al in the their compositions, and to a better adhesion of this scale to the steel substrates [10] and, hence, to significantly reduced breakaway. The mentioned point indicated for steels can be also applied for the aluminized coatings. Moreover, according to some works conducted by Perez et al and Aguero et al [11, 12], the high temperature steam oxidation test results confirm the benefits obtained with the aluminide-based coatings. It should be outlined that, in the studied coatings, the major constituents are presented as intermetallides $Fe(Cr,Ni)_xAl_y$, and, because of that, the "activity" of the elements, such as Fe, Cr and some others, in the intermetallides and the scale growth rate are reduced compared to the case when these elements are presented in bare steels.

Some authors [13, 14] suggest failure of stainless steels at high temperature steam conditions with excessive growth of the chromia scale through its cracking and buckling due to the influence of hydrogen formed during steam oxidation. Thus, they assume that hydrogen penetrates through the scale accumulating in pores with creation of mechanical stresses at high temperatures with buckling leading to accelerated corrosion degradation. The hydrogen effect is omitted in this work since this effect is expected in steam when steam dissociation occurs at high temperatures and high pressures (e.g. when pressures significantly exceed 4-5 MPa). Comparing bare metals tested and the aluminized coatings with sufficient thicknesses, it could be expected, according to the works [15, 16, 17], that the latter ones would significantly reduce the hydrogen permeation not only due to sufficient thickness of rather inert and well consolidated aluminides, but also due to alumina film formation on the coating surface. The data of effectiveness of the BN coating on stainless steel against hydrogen permeation reported by Tamura et al [18] suggest to assume that the coating E (aluminized with a thin BN layer) may be also a promising solution for the steel protection against high temperature oxidation when hydrogen may occur.

In the case of the uncoated 800H and, especially, 310SS after a long time exposure at 800 ^{o}C, the chromium concentration in the steel near the scale interface is still sufficient to form the protective chromia scale. However, in this study, some nodule formation was observed for 310SS that may be dealt with a local reduction of the chromium content and $MnCr_2O_4$ spinel formation. In this case, the integrity and bonding of the oxide scale become lower resulting in its breaking and detachment. This leads to the quicker oxidation of iron and nodule formation creating the surface structure similarly to the data presented earlier [19, 20]. This structure basically contains two layers where the outer layer is iron-rich with a small quantity of chromium, whereas the inner layer consists of mixture of Cr, Ni and Fe oxides or/and spinel-type compounds, like $FeCr_2O_4$ and $(Ni,Fe,Cr)_3O_4$, which are not protective [20], and the nodules underneath. The nodule formation on the surface of bare 800H was negligible. However, the contents of Cr and Ni on the surface of 800H varied from ~2 to 40-45 wt.-% and from 6 to 30-33 wt.-%, respectively.

Finally, the phenomenon of the Cr volatilization that may be considered in this study is questionable. In general, Cr evaporation occurs in stainless steels and Cr-contained alloys through the formation of volatile chromium oxyhydroxide $CrO_2(OH)_2$. However, according to the study conducted by Young and Pint [21], volatile chromium oxyhydroxide $CrO_2(OH)_2$ particularly occurs at high oxygen partial pressures (much greater than 1 bar), however, in the present work, the long-time testing procedure was conducted in the water steam (not in pure oxygen where oxidation rate would be greater) at the pressure of 1 bar only, that is too low to induce evaporation process of the Cr-based species. The formation of $MnCr_2O_4$ spinel (observed by XRD in other study where the same 310SS were used in the same conditions [3]) should significantly reduce the probability of the Cr-oxyhydroxide formation [22]. In this study, no "straightforward" conclusion about the Cr-evaporation could be made; however, more specific investigation may be undertaken. As opposed to uncoated steels, the presence of rather thick (>100 μm) aluminized coatings totally eliminates the risk of evaporation of the Cr-based species from steels and alloys.

The studied coatings formed through the thermal diffusion technology can be manufactured on the industrial basis onto the components with different complex shapes and sizes, including onto inner or inner and outer surfaces of long tubing (up to 10 m length). Since the coatings' performance is related to the aluminides-based structure and by the formation of the inert oxide "skin" well-adhered to the main coating, these coating structures can be obtained also for many other steels and alloys. The formation of the multi-layered structures with different thickness of the protective layers can be managed by the process parameters and composition of the aluminizing process. The coatings can be recommended for the service in high-temperature steam oxidation, e.g. for coal-fired power plants, heat exchangers and many other related industrial applications.

Conclusions

Long-time testing of high-temperature steam oxidation was conducted to investigate the behaviour of the aluminized thermal diffusion coatings deposited on 310S and 800H substrates vs. bare materials. For the bare materials, the oxide Cr-oxide-based scales occurred, which have porous structures and weakly bonded to the substrate, that results in a high probability of the breakaway issue. The uncoated 800H substrate showed much higher oxidation rate compared to 310SS. A higher mass gain in 800H Incoloy was attributed mainly to lower Cr and higher Ni contents and it is confirmed by microstructure cross-section examination when 800H developed a much thicker oxide scale than that observed in 310SS.

The multi-layered coatings based on aluminides with substantial thicknesses (greater than 100 μm) demonstrated higher integrity after the exposure at the conducted testing conditions. No significant coating destruction, delamination and breakaway issue were observed. The improved protection is related to general inertness of aluminides and with a presence of a thin aluminium oxide-based film formed at high-temperature oxidation which also promotes the protection of steels from the occurrence and subsequent removal of Cr_2O_3-scales with a poor adhesion and with potentially elevated volatility. The diffusion-related bonding between the layers in the aluminized-based coating systems also promotes integrity at severe oxidation conditions. The hardness data results of the aluminide layers confirmed high integrity and reliable behaviour of these coatings. Additional top layers from inert ceramic materials (e.g. oxides of the metals of III-IV groups and BN) onto the main coating made of aluminides may be a positive step for improvement of oxidation resistance; however, it is important to avoid the breaking and destruction of this layer during the high temperature oxidation exposure.

The consideration of only mass change during the high-temperature corrosion testing does not provide the adequate information about materials' corrosion resistance. However, the conjunction of the assessment of the "kinetic" oxidation curves and structural and surface examination of the materials and hardness determination is the most effective route for the materials evaluation and prediction of their behaviour under corrosive environments.

References

[1] P. Ennis, W. Quadakkers, Implications of steam oxidation for the service life of high-strength martensitic steel components in high-temperature plant, International Journal of Pressure Vessels and Piping. **84(1/2)** (2007) 82.
[2] A. Fry, S. Osgerby and M. Wright, Oxidation of Alloys in Steam—A review, National Physical Laboratory (NPL), Teddington, United Kingdom (2002)
[3] T. Dudziak, V. Deodeshmukh, L. Backert, N. Sobczak, M. Witkowska, W. Ratuszek, K. Chruściel, A. Zieliński, J. Sobczak, G. Bruzda, Oxidation of Metals **87(1)**, 1398 (2016).
[4] V.P. Deodeshmukh, Oxidation of Metals, **79**, (2013), 567.
[5] S. Bsat, X. Huang, Oxidation of Metals, **84,** (2015) 621.
[6] H. Asteman, J.E. Svensson, L.G. Johansson, Oxidation of 310 Steel in H_2O/O_2 Mixtures at 600 ^{o}C: the Effect of Water-Vapour-Enhanced Chromium Evaporation, Corrosion Science, **44,** (2002) 2635.
[7] M. Lukaszewicz, N.J. Simms, T. Dudziak, J.R. Nicholls, Oxidation of Metals **79(5/6)** (2013) 473.
[8] X. Cheng, Z. Jiang, D. Wei, J. Zhao, B.J. Monaghan, R. J. Longbottom, L. Jiang, Surface and Coating Technology, **258,** (2014) 257.
[9] X. Peng, J. Yan, Y. Zhou, F. Wang, Effect of Grain Refinement on the Resistance of 304 Stainless Steel to Breakaway Oxidation in Wet Air, Acta Materiala, **53(19),** (2005) 5079.
[10] Z. Yu, M. Chen, C. Shen, S. Zhu, F. Wang, Corrosion Science, **121,** (2017) 105.
[11] F.J. Perez, S.I. Castaneda, Surface and Coatings Technology, **201** (2007) 6239.
[12] A. Aguero, R. Muelas, A. Pastor, S. Osgerby, Surface and Coatings Technology, **200** (2005) 1219.
[13] F. H. Stott F. I. Wei, C. A. Enahoro, Materials and Corrosion, **40** (1989) 198.
[14] J. Yuan, W. Wang, H. Zhang, L. Zhu, S. Zhu, F. Wang, Corrosion Science, **109** (2016) 36.
[15] C.H. Henager, Hydrogen Permeation Barrier Coatings, pp. 181-190 in Materials for Hydrogen Economy, Chapter 8 (Eds.: R.H. Jones and G.J. Thomas), CRC Press, 2007
[16] K.S. Forcey, D.K. Ross, C.H. Wu, Journal of Nuclear Materials, **182** (1991) 36.
[17] H. Glasbrenner, A. Rerujo, E. Serra, Journal of Fusion Technology, **28 3P2** (1995), 1159.
[18] M. Tamura, M. Noma, M. Yamashita, Surface and Coatings Technology, **260** (2014) 148.
[19] S.R.J. Saunders, M. Monteiro, F. Rizzo, Progress in Material Science, **53** (2008) 775.
[20] X. Cheng, Z. Jiang, D. Wei, J. Zhao, B.J. Monaghan, R. J. Longbottom, L. Jiang, Surface and Coatings Technology, **258** (2014) 257.
[21] D.J. Young, B.A. Pint, Oxidation of Metals, **66, 3/4** (2006) 137.
[22] G. Holcomb, D.E. Alman, Scripta Materiala, **54** (2006) 1821.

Selected carboxylic acids as active corrosion inhibitors for zinc metal in HCl

Lutendo Chester MURULANA[1], Tshimangadzo NESANE[1]

[1] University of Venda, South Africa

In this investigation, two selected carboxylic acid derivatives namely, 3-phenyl amino propanoic acid (PAPA) and 3-phenoxy propanoic acid (POPA) have been synthesized and evaluated as corrosion inhibitors for zinc in hydrochloric acid solutions utilizing experimental techniques such as gravimetric analysis, electrochemical impedance spectroscopy (EIS), potentiodynamic polarization (PDP) as well as theoretical method (using DFT). These organic synthesized corrosion inhibitors were characterized using Nuclear magnetic resonance (NMR) and Fourier transform infrared (FTIR) spectral analyses. The interactions between the inhibitor molecules and metallic surfaces were further investigated utilizing scanning electron microscopy (SEM) and Fourier transform infrared spectroscopy (FTIR).

These compounds were found to protect zinc through spontaneous adsorption on to zinc surface, thereby blocking the active sites. The results obtained from the gravimetric analysis indicate that the corrosion inhibition efficiency of zinc metal in acidic environment increased with an increase in the concentration of the inhibitors. The results also suggest that PAPA served as a better inhibitor in comparison to POPA with an inhibition efficiency of 89.22% at the concentration of 12 x 10^{-3} M. Thermodynamic and kinetic parameters such as ΔG^{o}_{ads}, E_a, ΔS^* and ΔH^* were calculated from the data collected and the values of E_a and ΔH^* suggested a mixed-type adsorption process as seen from the Langmuir isotherms.

Results from electrochemical impedance and potentiodynamic polarization showed exceptional correlation and revealed that these compounds are mixed-type inhibitors. Quantum chemical calculations and quantitative structure activity relationship (QSAR) were also utilized to investigate the preferred reactive sites on the molecules of the utilized carboxylic acids and to correlate the experimentally determined inhibition efficiency with the theoretically determined inhibition efficiency. The results indicate the preferred sites for interaction between the inhibition and metal surface would be the aromatic ring as well as the heteroatoms present in each compound.

Keywords: Zinc, corrosion inhibition efficiency, adsorption, carboxylic acids, electrochemical techniques

Effect of Alloyed Nb on the Air Oxidation Behaviour of Ni-Cr-Fe Alloys at 1000 °C

Yaxin XU[1], Wenya LI[1]

[1] *Northwestern Polytechnical University, China PR*

Ni-25Cr alloys alloyed with Fe (5 wt.% and 15wt.%) and Nb (0, 1 and 2 wt.%) were designed and their static oxidation behaviour was evaluated at 1000 °C up to 100 h. The mass gain of alloys during corrosion test was recorded to determine oxidation kinetics, and the morphology and structure of oxide scales after oxidation were characterised by SEM, XRD and EPMA. Results reveal that alloyed Nb has a positive impact on oxidation in both the kinetics and oxide scale structures. The parabolic rate constant k_p of Nb-alloyed Ni-25Cr-15Fe alloys are nearly one-fourth the value for Nb-free alloy. The oxide scales change from spinel-included multi-layers in the Nb-free alloy to a much thinner Cr_2O_3 single-layer in the Nb-alloyed alloys. NbC precipitates and slightly dissolved Nb account for these changes.

The microstructure characterization of the oxide scale created on Inconel 686 clad

Damian KOCLĘGA[1], Agnieszka RADZISZEWSKA[1], Axel KRANZAMNN[2], Stanisław DYMEK[3]

[1] *Agh University of Science and Technology, Poland*

[2] *BAM - Federal Institute for Materials Research and Testing, Germany*

[3] *AGH University of Science and Technolgy, Poland*

The work presents the microstructure and chemical composition of the oxide scales created on the clad from Ni - base alloys after corrosion experiments in aggressive gases and ashes atmospheres. The clad from Inconel 686 on the low carbon steel 13CrMo4-5 were performed by QS Nd:YAG laser cladding process. The Ni-Cr-Mo-W alloy is characterize by the excellent high temperature corrosion resistance, good strength and good ability to work in aggressive environments. The experiment were performed in oxidation gases atmosphere like O_2, CO_x , SO_x. The second variation of corrosion experiments were performed in reduction atmosphere with ashes contained elements like: Na, Cl, Ca, Si, C, Fe, Al etc. After 240h and 1000h corrosion experiment in both cases the oxide scale on the substrate and clad were created. The sulfur compounds were found on the top of clad surface (EPMA) while after the ashes experiments the investigation revealed the presents the higher contents of silica compounds on the top of clad surface. The microstructure, chemical composition of the clad and scales were investigated by means of a light microscope, the electron microscopes (SEM, TEM) equipped with the EDS detectors.

Corrosion Mitigation of Fire Sprinkler Pipes under Nitrogen Supervision

Paul SU[1], **Jens ALKEMPER**[1]

[1] *FM Global, United States of America*

Corrosion Mitigation of Fire Sprinkler Pipes under Nitrogen Supervision

Paul Su* and Jens Alkemper

FM Global Research, Norwood MA, 02062 USA

* Corresponding author:

For steel sprinkler pipes, corrosion that results in pipe leakage or obstruction is the most significant issue for owners of water-based fire protection systems (FPS), in terms of both cost and system reliability. The buildup of insoluble deposits inside steel sprinkler piping (e.g., tubercles and pipe scale due to corrosion, calcium carbonate and magnesium carbonate deposits due to water hardness) can obstruct water flow in the pipe and/or plug sprinklers. Repair of corrosion damage due to mechanical failure or water damage significantly adds to the lifecycle cost of sprinkler protection systems in buildings. Hence, to properly maintain functionality of FPS, corrosion mitigation of sprinkler pipes is needed.[i],[ii]

For dry pipe systems (dry and preaction), residual water and dissolved oxygen in water are the main factors contributing to corrosion and leakage of sprinkler piping. Corrosion mitigation of steel sprinkler piping in these systems has been successfully achieved through nitrogen supervision to reduce oxygen concentration by using an on-site nitrogen generator.[iv],[v]

For wet pipe systems, causes for pipe corrosion and leakage are more complicated than those for dry pipe systems. Common causes may include trapped air, corrosive water chemistry, periodically supplying oxygenated water into the FPS (fresh sprinkler water recharged during regular maintenance), and microbiologically influenced corrosion (MIC).[1,][vi]

This paper presents the corrosion mitigation studies for dry and wet pipe systems filled with non-corrosive tap water and/or untreated pond water containing corrosive anions and MIC-causing bacteria, under nitrogen supervision. The results showed that pipe corrosion in both dry and wet types of FPS can be mitigated by using nitrogen as a supervision gas.

Keywords: Sprinkler Pipe Corrosion, Nitrogen Purging, Corrosion Mitigation, Dry, Preaction and Wet Fire Protection System, MIC

[i] P. Su and D. B. Fuller, "Corrosion and Corrosion Mitigation in Fire Protection Systems," FM Global technical report, 2nd Edition, July 2014.

[ii] P. Su, F. Tatar, S. Chivukula, and W. W. Doerr, "Weld Seam Corrosion of Steel Sprinkler Pipe," Corrosion 2013, Paper # 2140, NACE International, Houston, TX, 2013.

[iv] P. Su, S. Chivukula, F. W. Tatar, and B. Swinnerton, "Corrosion of Sprinkler Piping Under Compressed Nitrogen and Air Supervision," Corrosion 2015, Paper # 5548, NACE International, Houston, TX, 2015.

[v] J. Tihen, "Corrosion Inhibition of Dry and Preaction Fire Suppression Systems Using Nitrogen Gas," 2016 NFPA meeting.

[vi] B. J. Little and J. S. Lee, "Microbiologically Influenced Corrosion," John Wiley and Sons, Inc., Hoboken, NJ, 2007.

High injection rates of corrosion inhibitor into wet gas-condensate pipelines, operational issues and consequences

Askar SOLTANI[1], Reza GHORBANI[2], Koorosh FARZANFAR[2], Reza JAN NESAR SHARGH[2]

[1] *SPGC(Soith Pars Gas Complex), Iran*
[2] *SPGC, Iran*

Too much corrosion inhibitor injection is costly, adds no benefits, have environmental issues and can cause operational problems but too little can increase the risk of the corrosion attack. So The concentration of residual corrosion inhibitor monitoring is a common solution which oil and gas operators use to monitor the level of protection in transmission wet gas-condensate pipelines from offshore to onshore facilities as well as the monitoring of the corrosion rate of the coupons which installed at the end of the pipeline in onshore area. In this paper , the main purpose is to focus on the origin of the deposits which observed on the tubes of stabilization reboiler and decreased its thermal efficiency and led to periodic unwanted shut downs in order to remove the deposits. Raw condensate, gas and glycolated water mixture comes from reception unit to stabilization unit and after separation this raw condensate will be introduced into the stabilization column. There is a kettle type heat exchanger called stabilizer reboiler in order to increase the efficiency of the stabilization column. The raw condensate contains some water phase which there is some amounts of residual corrosion inhibitor injected into the feedstock transmission pipelines at offshore platforms. Also there are some corrosion inhibitor injection points in stabilization unit. Observed deposit on the tubes was a black, sticky with gritty texture solid phase which was surrounded by clear liquid phase. Both liquid and solid phases were analyzed by gas chromatography technique. Gas chromatography results indicated that the liquid phase was hydrocarbon condensate. Black and gritty solid phase put into glycol and then was filtered. Liquid and solid phases were separated after filtration and gas chromatography were done on them.EDX analysis on the solid sample showed that it mainly contains iron sulphide and salts (e.g. KCl and NaCl).Heating the liquid phase yielded a highly viscous, sticky and golden brown liquid which was identical to polyamide resins from the appearance and texture point of views. The good agreement between the peaks of gas chromatography on the actual corrosion inhibitor and liquid phase of the deposit which was put into the glycol confirmed that one of the major constituents of the deposit is remnant corrosion inhibitor as the active component which can leave behind when the main solvent has evaporated off in the reboiler. This component acts as glue and adhe re corrosion products and salts together and form a sticky and gritty bulk which accumulates on the tubes of the reboiler and decrease its thermal efficiency. Laboratory analyzing revealed that there were very high amounts of residual inhibitor at glycolated water phase at reception unit. After decreasing the injection dosage of corrosion inhibitor into the transmission pipeline and controlling the dosage rate of the corrosion inhibitor pumps in the stabilization unit, the problem was cleared.

The Effect of Gd and Ce on the Cyclic Oxidation of Ni-Base Superalloy

Dae Won YUN[1], **Young Soo YOO**[1], **Hi Won JEONG**[1], **Seong Moon SEO**[1]

[1] *Korea Institute of Materials Science, Korea, South*

Elements with high oxygen affinity, which are called reactive elements, are often added to a high temperature alloys to improve oxidation resistance. Reactive elements are known to promote selective oxidation of Cr and Al and to suppress formation of transient oxides. The addition of reactive elements can be more effective in enhancing oxidation resistance under a cyclic condition by increasing the adhesion strength between an oxide scale and a metal substrate. Several mechanisms are proposed to explain the beneficial effects of reactive elements. They can be summarized as follows. Firstly, small oxide particles of reactive elements segregate at grain boundaries of oxide scales and alter the diffusion along the grain boundaries, which are a fast diffusion path. Secondly, oxides of reactive elements segregate at the interface between an oxide scale and a metal substrate and prevent the annihilation of cation vacancies at the interface. Thirdly, oxide of reactive elements can remove the impurities such as sulfur, which are detrimental to adhesion of oxide scales, from the metal substrate by absorbing them. In practice, mostly La and Y are alloyed as a reactive element. In some cases, Ce is also used as a reactive element.

In this study, the effect of Gd and Ce addition up to 0.02 wt% on cyclic oxidation of an alumina-forming Ni-base superalloy at 1100°C has been investigated. The samples were exposed at 1100°C for 1 h and cooled down to room temperature for 30 min repeatedly up to 50 cycles. After the cyclic oxidation experiment, the microstructure of oxide scale was observed by scanning electron microscope. Secondary ion mass spectroscopy depth profiling was utilized to identify the distribution of each element in the oxide scale. It was found that Gd as well as Ce has beneficial effect on the oxidation resistance of alumina-forming Ni-base superalloys.

Increasing acid dew point corrosion resistance by thin coating layer on steel

Changhoon CHOI[1], Byoung Ho LEE[1], Minho JO[1], Myung Soo KIM[1]

[1] *POSCO, Korea, South*

Acid dew point corrosion is a common phenomenon for industrial equipments such as duct, stack, gas gas heater (GGH) in plants using fossil fuels. During the burning process, sulfur in fossil fuel is oxidized to produce sulfuric dioxide (SO_2) and further oxidized to sulfuric trioxide (SO_3). Under the dew point temperature, water drops form on the surface of equipments along with sulfuric trioxide to develop sulfuric acid (H_2O+SO_3àH_2SO_4). Special steel products that are strong against acid dew point corrosion are on the market as cold rolled (CR) or hot rolled (HR) steel. Although these products have stronger corrosion resistance than normal steel, life span is still short and further increase of corrosion resistance is required. In this research, we applied thin coating layer to steel and studied how corrosion resistance is changed. Different types of coating were applied and tested. Localized corrosion that could occur by surface coating was monitored and possible solution was proposed. Coating layer was observed by various analytical tools including FT-IR, SEM, EPMA, GDS etc.

Electrochemical evaluation of volatile corrosion inhibitors in the presence of condensation water

Maximilian SIEBER[1], **Signe LAUTNER[2]**, **Frank FAßBENDER[3]**

[1]EXCOR Korrosionsforschung GmbH, Dresden, Germany, maximilian.sieber@excor.de
[2]ibd., signe.lautner@excor.de
[3]ibd., frank.fassbender@excor.de

Abstract: An appropriate corrosion protection is required for transport and storage of high-quality goods and semi-finished parts made from metals. The application of volatile corrosion inhibitors (VCI) as a means of temporary corrosion protection is based on the release of corrosion inhibitors from films, papers, diffusers or oils to the gaseous phase. By adsorption on the metal surface, the inhibitors prevent the degradation of the primary oxide layer and hence impede corrosion. In principle, the anticorrosive effect is independent of the part geometry, because the protected surfaces are not in direct contact with the VCI material. Further, the VCI method does not require costly cleaning of parts prior to further use after transport and storage.
A new approach for the evaluation of VCI is presented. A metallic sample is exposed in a chamber filled with air along with a VCI-containing product. In the course of the test, the air is humidified and the sample is then cooled down below the dew point, which leads to condensation of humidity. Electrochemical DC and AC methods, namely polarization resistance measurements and electrochemical impedance spectroscopy, are applied to characterize the state of the metallic surface under the condensed water film. The time-resolved analysis of the measurements provides a basis for the quantitative evaluation of the VCI´s anticorrosive effect or a possible corrosion attack.
The present work focusses on establishing the interrelations between the electrochemical response of the system and the test regime including initial exposition time of the dry sample along with the VCI in the test volume, the amount of VCI and the dwelling time of the sample under condensed humidity. The validity of the measurements is verified and the range of deviation is estimated. Exemplarily, the anticorrosive effect of various VCI-containing products on mild steel is evaluated. In summary, the presented application-oriented test is proven as a suitable method for the quantified assessment of VCI-containing products.

Keywords: Volatile Corrosion Inhibitors; Dew Point; Condensation; Electrochemical Impedance Spectroscopy; Polarization Resistance

Introduction

Problem statement and proposed solution

Bare metal surfaces are prone to corrosion not only during service, but already during transportation and storage. Corrosion may be induced by as little as a temperature change and the inevitable condensation of humidity from the atmosphere on the metal surface. Temporary corrosion protection is one possibility to protect semi-finished parts and assemblies against corrosion. A common means is the application of volatile corrosion inhibitors (VCI), which is based on the release of corrosion-inhibiting substances from films, papers, diffusers or oils to the gaseous phase within a packaging. By adsorption on the metal surface, the inhibitors prevent the degradation of the primary oxide layer and hence impede corrosion. The main advantages of this method are: (1) the anticorrosive effect is independent of the parts´ geometry, because no direct contact of the parts´ surfaces with the VCI material is required – exposure to the same volume of air within a packaging is sufficient; (2) additional process steps before packing like oiling are not necessary; (3) no costly cleaning of the parts prior to further use after transport or storage is required and (4) disposal of any cleaning baths after removal is avoided, because the inhibitors volatilize within a short period of time after unpacking of the parts. For further information on the VCI-method see [1–5].

Both the development of new VCI-containing products and the quality control of commercially available VCI-containing products require appropriate test methods to evaluate their corrosion-inhibition effects. To date, a plethora of tests have been developed, which allow for qualitative and/or quantitative assessment of VCI-containing products. These methods may be categorized as follows:

(1) methods which are based on the chemical analysis
   a. of the VCI-containing products themselves or
   b. of a well-defined volume of air in which the VCI-containing product is exposed or
   c. of a well-defined sample surface which is exposed to the VCI-containing product in a volume of air

(2) methods which are based on the exposure of a metallic sample to a climatic stress, i.e. an atmosphere with well-defined temperature and humidity,
   a. within a packaging consisting of VCI-containing products (e.g. film, paper, goods carrier)
   b. within a packaging along with VCI-containing products (e.g. oil, diffuser)

(3) methods which are based on electrochemical measurements on a metallic sample in an aqueous solution, which
   a. includes the VCI-containing products or
   b. is obtained from the VCI-containing products by elution of the latter

The methods of category (1) are advantageous, because they provide quantitative results with regard to the amount of corrosion inhibitors, which are contained within the products and/or which are set free to the surrounding atmosphere. However, these methods are generally costly both in terms of the time to carry out the analyses and in terms of the required equipment and expertise. The approaches of category (2) give valid results about the corrosion-inhibition effect of any VCI-containing products under the respective test condition, but almost inevitably include a subjective visual assessment of the sample surfaces after the test, which makes them prone to errors and/or significant deviations when carried out by different workers or in different institutions. The methods (3) provide quantitative information about the corrosion-inhibition effect of VCI-containing products. Anyhow, the results obtained with this approach are actually often misleading, because the mode of action of inhibitors cannot be assumed to be identical in the test medium (aqueous solution) and under atmospheric conditions, which normally apply during service.

It has therefore been the aim of this study to propose a test method, which provides objective results on the corrosion-inhibition effect of VCI-containing products in a comparatively short amount of time. The approach includes the exposure of a metallic sample surface alongside with a VCI-containing product to a defined volume of atmosphere, the condensation of water on the sample surface by active cooling of the sample below the dew point and the characterization of the sample surface under the condensed water film by electrochemical methods, namely electrochemical impedance spectroscopy (EIS) and polarization resistance measurements. Consequently, (1) the release of inhibitors from the VCI-containing product, (2) the transport of these inhibitors to the metallic surface in the gaseous phase, (3) the interaction of the inhibitors with the primary oxide layer on the surface and (4) the ability to impede corrosive action under climatic stress contribute to the result of the phenomenological test. Thus, with the VCI mode of action simulating the conditions during service, the test shall allow for a quantitative assessment of the state of the sample surface and hence the corrosion-inhibition effects of VCI-containing products.

Correlation between sample corrosion and results of electrochemical measurements

Numerous researchers have established interrelations between the state of a passive or corroding metal surface and the parameters obtained from the electrochemical characterization of the same surface. When utilizing direct current (DC) methods for electrochemical characterization, a state of high surface activity is generally reflected by higher current density under a certain polarization, which corresponds to a lower polarization resistance. If such behavior is recognized in an equilibrium state or near-equilibrium state of the corrosion system, it almost certainly indicates corrosive action. The formation of a passive state is an exception, because in this case higher current densities can be registered, but only for a short period of time (non-equilibrium). In the passive state, the current density remains at low values (high polarization resistance). When utilizing alternating current (AC) for electrochemical characterization, namely electrochemical impedance spectroscopy (EIS), the obtained result, i.e. the impedance at a certain frequency, includes an information about the amplitude ratio of the sine-wave potential/current signals and the phase shift between said signals. One possible method to derive quantitative information about the surface state is to model the frequency-dependent impedance with equivalent circuits containing resistive, capacitive, inductive and other elements. One popular equivalent circuit is so-called Randles circuit [6], which includes a capacitance reflecting the behavior of the double layer at the substrate/electrolyte interface and an impedance element, e.g. a simple resistance, which accounts for the electrochemical reaction at the said interface. With regard to corrosion activity, it has been shown in [7–10] that higher values of the capacitance and lower values of the resistance correspond to a corrosion-active surface state. Contrariwise, lower values of the capacitance and higher values of the resistance correspond to a passive surface state. In the present study, these already established interrelations between surface state and parameters obtained from electrochemical measurements are utilized for the assessment of VCI-containing products.

Experimental Details

Materials and preparation

A metallic sample was cut from a rod of steel S235JRG, the nominal composition of which is given in Table 1. The sample was 16 mm in diameter and approx. 10 mm in height. Before each test, one of the faces of the sample was grinded with SiC-paper (600 grit), rinsed with de-ionized water, rinsed with ethanol and dried under a stream of warm air. For the electrochemical measurements under a condensed water film, the sample was placed in a sample holder made from polyamide. In the sample holder, there was a defined aperture, under which the grinded surface was positioned. There, a ring seal was inserted between the sample holder and the sample. Above the aperture, two auxiliary electrodes were positioned, which were used as counter electrode (platinized titanium mesh) and reference electrode (platinum wire) for the electrochemical measurements. The principal layout of the sample holder and the relevant dimensions are given in Figure 1. The sample holder was placed in a chamber made from PMMA (tube) and POM (lid) with a volume of 3.5 l. A peltier element was mounted to the bottom of the chamber. In the chamber, there was a profile made from aluminium with an epoxy coating. This profile was thermally connected to the peltier element. The sample holder was placed under the profile and was also thermally connected to the peltier element so that small drops of water, which condensed on the auxiliary profile, could support the wetting of the sample surface. Additionally, there was a water reservoir with a volume of approx. 50 ml at the bottom of the chamber, from which the atmosphere in the chamber was humidified by evaporation of the water. The layout of the test chamber is given in Figure 2.

Along with the sample holder, a diffuser unit of a commercially available, nitrite-free VCI material was placed in the chamber. The chamber was sealed so that no exchange with the surrounding atmosphere was possible. The test procedure was as follows:

(1) conditioning: the sample and the VCI-diffuser were exposed to the atmosphere in the chamber (relative humidity of 60 % to 90 %); 90 min

(2) water film formation: the relative humidity in the chamber was increased to 100 % (supersaturation of the atmosphere with humidity) and the sample and the auxiliary profile were cooled below the dew point by the peltier element, which led to condensation of humidity from the atmosphere in the chamber and thus to formation of a film of water, which wetted the sample surface and the auxiliary electrodes; 45 min

(3) electrochemical measurement: the temperature of the sample was kept below the dew point during the measurements to avoid evaporation of the previously condensed film of water; 60 min

Figure 3 representatively shows the course of the relative humidity in the test chamber and the sample temperature relative to the dew point during a test. After the test, the samples were visually inspected utilizing a light microscope with digital image recording system (VHX-2000, Keyence, Japan).

Table 1 : Nominal composition of the sample material S235JRG

Element	C	Mn	P	S	Fe
Concentration in wt-%	<0.2	<1.4	<0.045	<0.045	balance

Electrochemical measurements

The electrochemical measurements were performed using a workstation IM6 (Zahner, Germany). The sample, the platinized titanium mesh and the platinum wire (see above) served as working, counter and reference electrode, respectively. Electrochemical impedance spectra were recorded in a frequency range of 20 mHz to 100 kHz with 4 frequency steps per decade using a potential amplitude of 10 mV. For reasons of stability, the potential around which the amplitude was applied was set so that the absolute current flow between counter electrode and working electrode was below 150 nA at any time. Each impedance value at a particular frequency was averaged from 11 periods for frequencies above 66 Hz and 3 periods for frequencies below 66 Hz. With the mentioned parameters, recording 1 spectrum required approx. 9 min of time. Several impedance spectra were recorded sequentially to obtain information on the state of the sample surface in dependence of the test duration. The recorded impedance spectra were fitted to a model with 2 time-constants (Figure 4) using Octave [11]. The value R_E, which represents the electrolyte resistance, was normalized before the fit. Unless stated otherwise, any specification of time in the context of electrochemical measurements is with reference to the start of the water film formation (phase 2 in Figure 3).

Additionally, a simple polarization measurement was carried out at the end of each test. With the contemporary free potential of the sample as reference point, anodic polarizations of 3 mV and 6 mV were applied and the current flow was measured. The polarization resistance R_{pol} was calculated using the following equation (with ΔE – change of potential, Δi – change of current density):

$$R_{pol} = \frac{\Delta E}{\Delta i} \qquad (1)$$

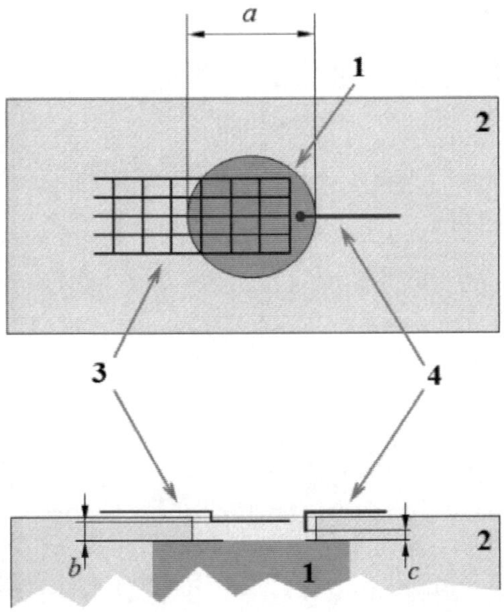

Figure 1: Principal layout of the sample holder (2) for electrochemical measurements of the sample (1) with the auxiliary electrodes (3 – platinized titanium mesh as counter electrode, 4 – platinum wire as pseudo-reference electrode). The dimensions are: a = 11.5 mm *(diameter of the active sample surface);* b = 2 mm *(distance between sample surface and counter electrode);* c = 1 mm *(distance between sample surface and tip of pseudo-reference electrode).*

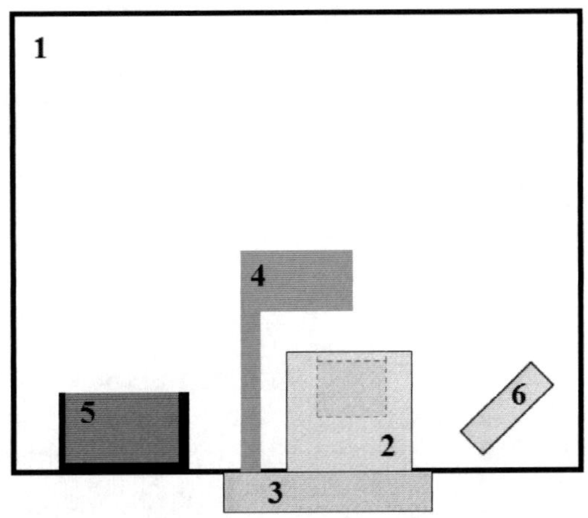

Figure 2: Principal layout of the test apparatus with the test chamber (1), in which the sample holder with the sample (2) is exposed along with the VCI-containing product (6) and water in a reservoir (5) for humidification of the atmosphere. At the bottom of the test chamber a peltier element (3) is mounted, which regulates the temperature of the sample. An auxiliary profile (4) is thermally coupled to the peltier element and supports the build-up of the water layer on the active sample surface.

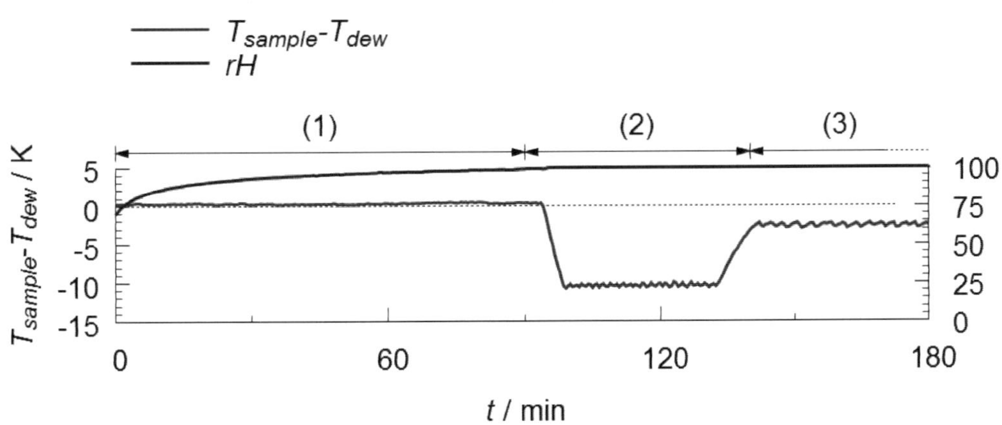

Figure 3: Representative course of the relative humidity and the difference between sample temperature and dew point during the test procedure with (1) – conditioning, (2) – water film formation, (3) electrochemical measurements. The dashed line marks the identity of dew point and sample temperature.

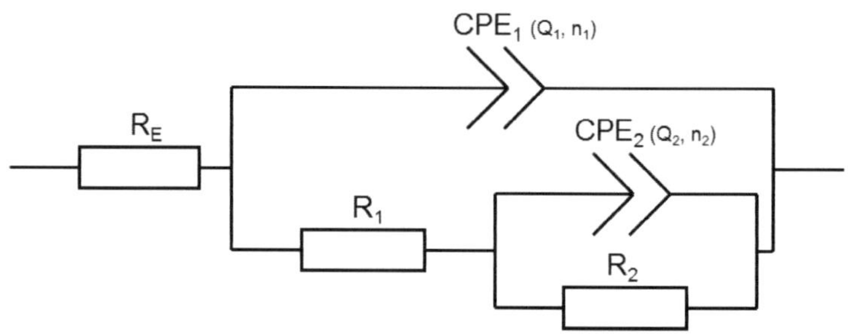

Figure 4 : Equivalent circuit with 2 time-constants for modelling the impedance data (the electrolyte resistance R_E is normalized).

Test variation

For the test conditions described above, the VCI diffuser impedes corrosion of the sample surface. In order to show the plausibility and principal capability of the test, the test was varied as follows to provoke a corrosion attack on the sample. For each of the variations, at least 4 samples were measured independently.

(1) Amount of VCI: The commercially available, nitrite-free VCI diffuser (1 unit) was stripped, so that only one quarter of its original mass remained (0.25 unit). Additionally, the tests were carried without any VCI-containing product.

(2) Conditioning time: The standard conditioning time (phase 1 in Figure 3) was 90 min under high humidity. The conditioning time was cut down to 0 min in order to weaken the interaction between the sample surface and the inhibitors released from 1 unit of VCI diffuser before the climatic stress of the sample.

(3) Dwelling time: The time span, for which the sample surface was covered with the film of condensed water (phase 3 in Figure 3) was prolonged for particular samples. The results of the electrochemical measurements were continuously checked. When the results indicated sample corrosion, the test was stopped and the sample surface was visually inspected.

Results and Discussion

Impedance behavior of the system and interpretation of equivalent circuit parameters

A representative impedance spectrum of a S235JRG sample under a water film formed by condensed humidity is shown in Figure 5. The system behavior is capacitive with 2 time-constants (2 loops in the Nyquist plot). The measured data is appropriately modelled by the chosen equivalent circuit. Preliminary tests have shown that the time-constant in the higher frequency range is originated in the test layout. The electrolyte, which is wetting the electrodes, has a very low electrical conductivity, because it practically consists of condensed humidity and possibly few amounts of corrosion inhibitors. Therefore, it acts as a dielectric between the electrodes and therefore produces a capacitive loop in the higher frequency range. When the electrical conductivity of the electrolyte is increased, e.g. by solution of corrosion products of a corroding sample or by tentative addition of a salt, the capacitive loop becomes less pronounced or vanishes completely. Also, when the geometry of the test layout is changed (distances between sample as working electrode and the auxiliary electrodes), the said capacitive loop also changes. In the equivalent circuit, the described capacitive behavior is represented by the elements R_1 and CPE_1 with pseudo-capacitance Q_1 and the exponent n_1. At lower frequencies, another capacitive loop is existent. It is related to the interface between the sample surface and the electrolyte and is represented by the elements R_2 and CPE_2 with pseudo-capacitance Q_2 and the exponent n_2. It has been established in numerous preliminary tests that for values Q_2 greater than approx. 10^{-4} $F \cdot s^{n-1} \cdot cm^{-2}$ corrosion occurs on the sample surface. Meanwhile, the value R_2, which is a measure for the charge transfer rate at the interface between sample surface and electrolyte, is not generally suitable as an indicator of corrosion. While high values of R_2 may indicate passivity, it has been found that especially for samples with heavy corrosion R_2 also increases, presumably because of the blocking of the surface and thus of the paths for charge transfer with corrosion products. However, for heavily corroded samples (e.g. without VCI-containing product in the test chamber) Q_2 increases significantly by orders of magnitude to values of 10^{-3} $F \cdot s^{n-1} \cdot cm^{-2}$ or even higher. Additionally, the deformation of the semicircle in the Nyquist plot modelled by R_2, Q_2 and n_2 changes, which is reflected by an increase of n_2 from values in the range of 0.7–0.9 (passive state or comparably weak corrosion attack) to values ranging from 0.9 to even greater than 1 (severe corrosion attack). While Kramers-Kronig relations [12] are generally fulfilled for the investigated system also for $n_2>1$, the interpretation of such behavior is arguable, because the validity of electrochemical impedance measurements during severe sample corrosion is contentious for the obvious violation of stability criteria, anyway. The observed behavior nonetheless allows for an objective discrimination between samples in the passive state and samples undergoing more or less severe corrosion by the performed EIS measurements. The weighting of the error function for fitting the equivalent circuit parameters to the measured EIS-data has been adjusted according to the findings of the preliminary tests in order to achieve maximum significance of the parameters R_2, Q_2 and n_2.

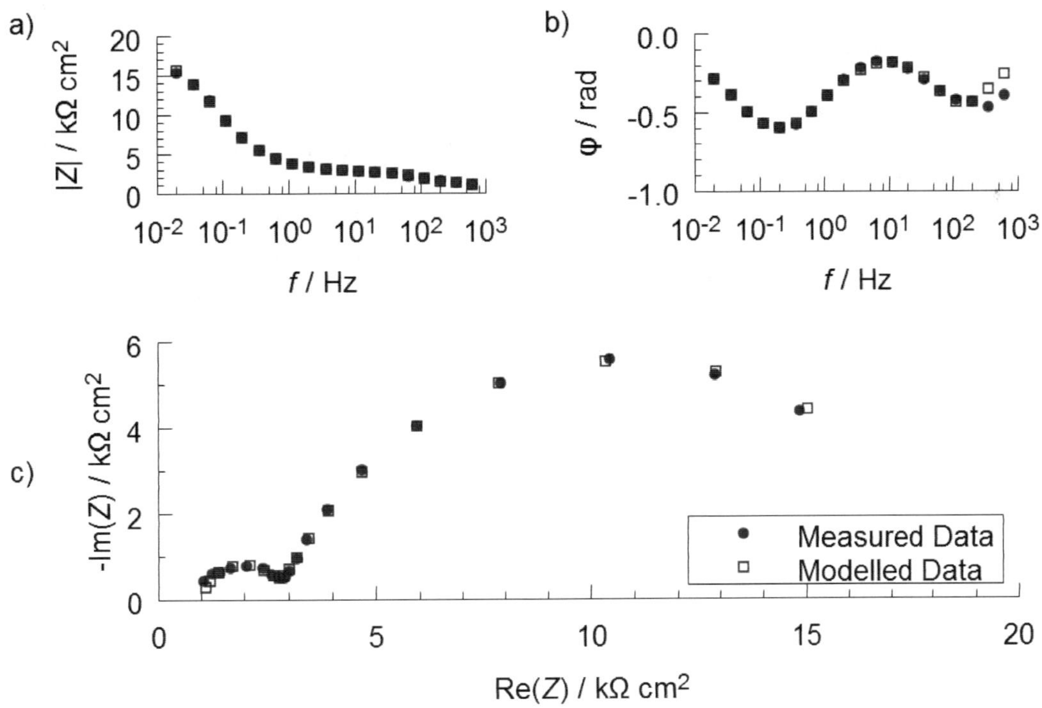

Figure 5 : Representative impedance spectrum obtained from a measurement of a S235JRG sample in the described testing device 90 min after the start of the water film formation in the presence of a commercially available, nitrite-free VCI diffuser; a – Bode plot of absolute value of the impedance over frequency, b – Bode plot of the phase shift over frequency, c – Nyquist plot of imaginary part of impedance over real part of impedance.

Factually, it has to be noted that the established limit ranges for the indication of corrosion have been narrowed down by correlating the equivalent circuit parameters with the occurrence of corrosion on the sample surface. Therefore, a relative deviation of a few percent for the particular values may occur when the method for fitting the equivalent circuit parameters to the measured data is changed (e.g. solver method, error function, weighting function etc.). For the given geometry, the relevant time-constant at lower frequencies has never been superimposed by the high-frequency time-constant.

<u>Influence of test variation on the inhibition effect and the obtained electrochemical parameters</u>

The parameter Q_2 has been established as a good indicator for corrosion activity on the sample surface in numerous preliminary tests. High values of the parameter represent corrosive attack. For evaluation of the samples, the maximum value of Q_2 obtained from the series of spectra recorded for each particular sample is chosen. The response of Q_2 to the variation of the test is shown in Figure 6.

Table 2 enlists the values representing the low-frequency time constant of the particular spectrum of each independently measured sample for which Q_2 is maximal. For the standard test procedure with 1 unit of the VCI diffuser and 90 min conditioning time, $\log_{10}(Q_{2max}$ in $F\cdot s^{n-1}\cdot cm^{-2})$ amounts to -3.98±0.06 for 5 independently investigated samples. In comparison, with the diffuser stripped down to ¼ of the mass of the previously used unit, $\log_{10}(Q_{2max}$ in $F\cdot s^{n-1}\cdot cm^{-2})$ amounts to -3.61±0.27 for 4 independently investigated samples. For the use of the original amount of VCI diffuser (1 unit) with a conditioning time of 0 min, $\log_{10}(Q_{2max}$ in $F\cdot s^{n-1}\cdot cm^{-2})$ is -3.94±0.09. Without any VCI-containing product in the test chamber, the pseudo-capacitance is again significantly larger with $\log_{10}(Q_{2max}$ in $F\cdot s^{n-1}\cdot cm^{-2}) = -3.09\pm0.27$.

The deviation within the measured values increases when corrosion attack takes place on the sample surface. This includes the deviations between independently measured samples as well as the deviations when the electrochemical parameters are obtained from a series of sequential measurements over time at any particular sample, especially when the corrosion attack is severe (e.g. without presence of VCI-containing material in the test chamber). Meanwhile, on the basis of the present results, the deviation is rather small when no or hardly any corrosion attack takes place. Interestingly, there is no statistically significant difference in Q_{2max} when 1 unit of VCI diffuser is exposed along with the sample in the test chamber with a conditioning time of 90 min or 0 min before the actual corrosion test. A t-test between the 2 groups with 1 unit of the VCI-diffuser yields a mean square between the groups of 0.0049 (effect of conditioning time variation) compared to a mean square within the groups of 0.0055 (error). Therefore, it is highly unlikely that the conditioning time has a significant influence under the considered test regime. Obviously, the commercially available nitrite-free VCI is able to immediately provide an inhibiting effect. This fact has also been proven by state-of-the-art test methods for VCI-containing products. When decreasing the amount of VCI from 1 unit to ¼ unit and to 0 (without VCI), there is a statistically significant change in the results with regard to Q_{2max}. An analysis of variance (ANOVA) of the 3 groups yields a mean square between the groups of 0.89 (effect of VCI amount variation) compared to a mean square within groups of 0.05 (error). The probability of erroneously considering the amount of VCI having a significant effect on Q_{2max} is 0.04 %, i.e. very low. In addition, t-tests between each of the pairs of the 3 groups (1 unit of VCI, ¼ unit of VCI, without VCI) show that each of the amounts of VCI results in a statistically significantly different value of Q_{2max} (values of mean squares and error probabilities are given in Table 3). The following conclusions are drawn from these analyses: (1) The commercially available VCI still has an inhibiting effect even when the amount exposed alongside with the sample in the chamber or the conditioning time is substantially reduced. (2) The test method is capable of appropriately discriminating different levels of corrosion-inhibiting effects.

The obtained values for the polarization resistance R_{pol} are depicted in Figure 7. In correspondence with the values for the pseudo-capacitance Q_2 and the visual appearance of the sample surface after the test, R_{pol} amounts to 59±11 kΩ·cm², 27±8 kΩ·cm², 41±13 kΩ·cm² and 18±4 kΩ·cm² for the tests with 1 unit of VCI diffuser and 90 min conditioning time, with ¼ unit of VCI diffuser and 90 min conditioning time, with 1 unit of VCI diffuser and 0 min conditioning time, and without VCI, respectively. Sample #2 without VCI was not considered because it is obviously an outlier with regard to R_{pol} (Table 2). In general, higher values for R_{pol} seem to correspond to lower values for Q_2 and vice versa, i.e. high values of R_{pol} represent low corrosive activity in accordance with the findings in the literature. The statistical analysis of the obtained results shows that the variation of the conditioning time (0 min, 90 min) with 1 unit of VCI diffuser does not have a clearly discriminable effect on R_{pol} with the mean square between groups being 684 (effect of conditioning time variation) compared to a mean square within groups of 141 (error). Therefore, the probability of erroneously considering the effect as statistically significant is 6.3 %, i.e. no statistically sound discrimination between the groups is possible, which is consistent with the results obtained for Q_{2max}. The reason for the comparatively high standard deviation of R_{pol} for a conditioning time of 0 min is supposed to originate from the fact that only for some of the samples a weak corrosive attack has been initiated until the end of the test. With regard to the amount of VCI (1 unit, ¼ unit, without), the ANOVA (Table 4) yields a statistically significant effect with a mean square between the groups of 1953 (effect of VCI amount) compared to a mean square within groups of 77 (error). Therefore, the probability of erroneously considering the effect of the VCI amount as significant is 0.02 %, i.e. it is highly likely for the VCI amount to affect R_{pol}. Unlike Q_{2max}, though, R_{pol} cannot serve to certainly discriminate between the particular amounts of VCI. While the t-tests between each of the pairs of the tested VCI amounts yield a statistically significant difference between 1 unit of VCI and both the other variants, it is shown that discrimination between the tests with ¼ unit of VCI and the tests without VCI is not statistically sound. Obviously, for both variants more or less considerable corrosive action takes place, which R_{pol} is not capable of discriminating between. Two main reasons are considered for the reduced informative value of R_{pol}: (1) In similarity to the equivalent circuit parameter R_2, R_{pol} may increase both in case of passivation and in case of the formation of a film of corrosion products on the sample surface. Therefore, values for R_{pol} generally have to be judged critically. (2) In this respect, it is noticeable that in the currently performed test procedure the polarization resistance has been obtained only once at the end of the test. It is expected that the value of R_{pol} decreases rather sharply when corrosion is initiated and only afterwards will rise again steadily if the sample corrodes to such an extent that substantial amounts of corrosion products are formed and block the surface. Whether or not the informative value of R_{pol} can be enhanced by sequential DC measurements over time will be subject to onward investigation.

Table 2: Polarization Resistance at the end of the test and equivalent circuit parameters at the time t_{Qmax} after the start of the water film formation on the sample, for which Q_2 is maximal, for each particular, independently measured sample.

VCI / conditioning time in min	#	R_{pol} in kΩ·cm²	$R_2(t_{Qmax})$ in kΩ·cm²	$\log_{10}(Q_2(t_{Qmax})$ in 10^{-4} $F \cdot s^{n-1} \cdot cm^{-2})$	$n_2(t_{Qmax})$	t_{Qmax} in min
1 unit / 90 min	1	65	28	-3.99	0.78	53
	2	65	23	-4.06	0.79	67
	3	49	93	-3.90	0.68	88
	4	70	25	-4.00	0.85	61
	5	45	22	-3.96	0.84	59
1 unit / 0 min	1	34	38	-3.80	0.66	78
	2	60	31	-3.97	0.67	50
	3	39	25	-4.00	0.85	61
	4	32	23	-3.97	0.78	75
0.25 unit / 90 min	1	38	27	-3.92	0.67	76
	2	27	21	-3.75	0.79	57
	3	21	179	-3.33	0.65	79
	4	23	39	-3.44	0.86	75
without VCI	1	15	11	-2.72	1.20	107
	2	230	56	-3.26	1.23	89
	3	22	23	-3.33	1.17	68
	4	15	387	-3.04	0.79	57

In order to further prove the validity and plausibility of the proposed test method, the results of the electrochemical characterization of the state of the sample surface are compared to the visual appearance of the sample surface after the test. Exemplarily, for sample #3 in the group with 1 unit of VCI diffuser in the chamber and a conditioning time of 90 min, Q_{2max} is shifted towards a value which indicates corrosion activity (Table 2). Visual assessment of the sample surface by light-microscopy shows that there are traces of corrosion (Figure 8). In contrast, for samples with Q_{2max} not exceeding 10^{-4} $F \cdot s^{n-1} \cdot cm^{-2}$, no signs of corrosion were observed. This certainly documents the capability of the electrochemical measurements to indicate corrosion. Similarly, the samples of the other groups with values of Q_{2max} just slightly exceeding the established corrosion-indicating limit of approx. 10^{-4} $F \cdot s^{n-1} \cdot cm^{-2}$ also show very few distinct corrosion spots, which are only identifiable when utilizing light-microscopy for surface assessment. When Q_{2max} increases further – approx. $\log_{10}(Q_{2max}$ in $F \cdot s^{n-1} \cdot cm^{-2}) \geq -3.8$ – corrosion of the sample surface is recognizable for the unaided eye.

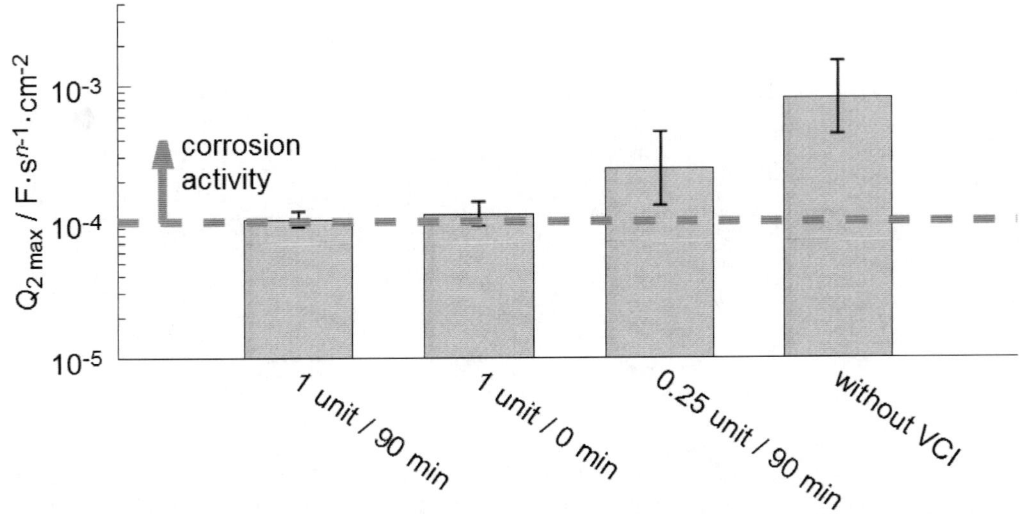

Figure 6: Maximum value of pseudo-capacitance Q_2 (mean values and standard deviations calculated from logarithmized values) obtained from the EIS measurements approx. 45–120 min after the start of the water film formation (phase 2) for different amounts of VCI (1 unit, ¼ unit, without VCI) and for different conditioning times (0 min, 90 min); values greater than approx. 10^{-4} $F \cdot s^{n-1} \cdot cm^{-2}$ indicate corrosion activity at the sample surface.

Table 3: Mean squares between groups (msq_x), mean squares within groups (msq_e) and probability of erroneously considering the effect by variation between the groups on Q_{2max} as statistically significant (p_{err}) as obtained from t-tests for each pair of the 3 amounts of VCI (conditioning time 90 min). The preliminarily performed ANOVA yields a high likeliness of Q_{2max} being significantly affected by the amount of VCI exposed along with the sample in the test chamber. The results of the t-tests show that each variation of the amount of VCI yields a statistically significant effect on Q_{2max}.

Amount of VCI	¼ unit	without
1 unit	$msq_x = 0.31$ $msq_e = 0.03$ $p_{err} = 2.0$ %	$msq_x = 1.78$ $msq_e = 0.03$ $p_{err} < 0.1$ %
¼ unit		$msq_x = 0.55$ $msq_e = 0.07$ $p_{err} = 3.6$ %

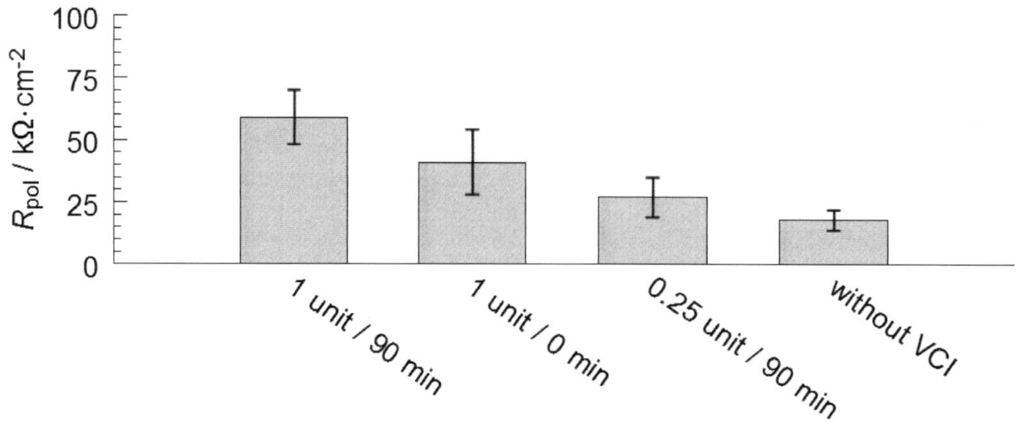

Figure 7: Polarization resistance R_{pol} as a measure for corrosion activity (low values represent high corrosion activity) obtained from polarization measurement at the end of each test for different amounts of VCI (1 unit, ¼ unit, without VCI) and for different conditioning times (0 min, 90 min).

Table 4: Mean squares between groups (msq_x), mean squares within groups (msq_e) and probability of erroneously considering the effect by variation between the groups on R_{pol} as statistically significant (p_{err}) as obtained from t-tests for each pair of the 3 amounts of VCI (conditioning time 90 min). The preliminarily performed ANOVA yields a high likeliness of R_{pol} being significantly affected by the amount of VCI exposed along with the sample in the test chamber. The results of the t-tests show that by considering R_{pol} it is not statistically safe to discriminate between a weak corrosion-inhibiting effect (¼ unit of VCI) and no corrosion-inhibiting effect (without VCI), while the corrosion-inhibiting effect provided by 1 unit of VCI is clearly discriminable.

Amount of VCI	¼ unit	without
1 unit	msq_x = 2212 msq_e = 95 p_{err} = 0.2 %	msq_x = 3224 msq_e = 87 p_{err} < 0.1 %
¼ unit		msq_x = 169 msq_e = 41 p_{err} = 9.9 %

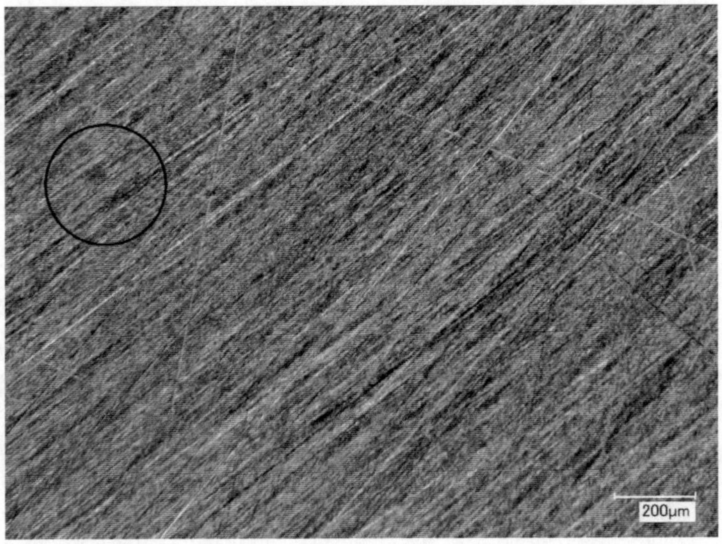

Figure 8: Surface appearance of sample #3 with 1 unit of VCI diffuser and 90 min conditioning time after the test. The black circle indicates a local corrosion attack. Further, traces of surface corrosion are visible in form of a slight brownish discoloration of the area around the indicated corrosion spot.

The peltier element at the bottom of the test chamber allows for the variation of the temperature of the sample with respect to the dew point. The results of an EIS measurement during a prolonged test with a temperature cycle after the formation of the water film is shown in Figure 9. After the conditioning time and the formation of the water film on the sample surface the sample is cyclically heated and cooled. In the present case the film of water on the sample surface stays intact so that wetting of the auxiliary electrodes is ensured at all times. Thus, continuous electrochemical measurements are possible during the whole test. After the formation of the water film, the pseudo-capacitance Q_2 is significantly below 10^{-4} F·s^{n-1}·cm^{-2}. During the 6th temperature cycle Q_2 increases sharply, which indicates corrosion activity at the sample surface. The visual appearance of the sample surface after the end of the test is shown in Figure 10. A slight brownish discoloration and few local corrosion spots are visible. However, no signs of heavy corrosion are existent. Thus, it is plausible that the corrosion has indeed been initiated during the last temperature cycle as indicated by the electrochemical measurement.

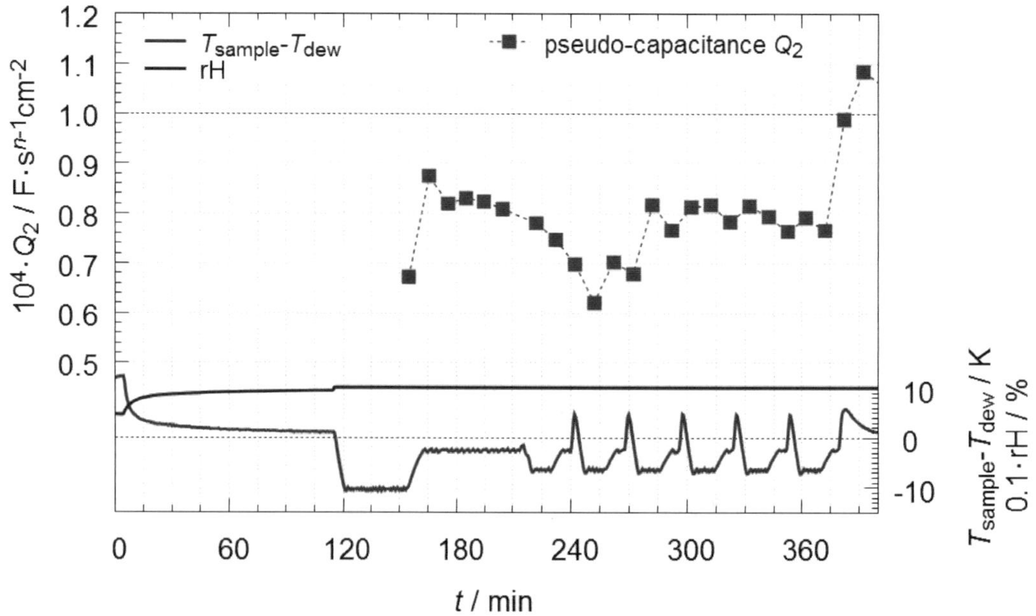

Figure 9: Pseudo-capacitance Q_2 obtained from an EIS measurement during a prolonged test with a temperature cycle after the formation of the water film on the sample surface; for reference, the difference between sample temperature and dew point as well as the relative humidity in the test chamber are shown in the bottom section of the diagram; during the 6th temperature cycle, the value of Q_2 increases sharply and becomes greater than 10^{-4} F·s^{n-1}·cm^{-2}, which indicates corrosion activity on the sample surface; the result of the visual inspection of the sample surface immediately after the last cycle is shown in Figure 10.

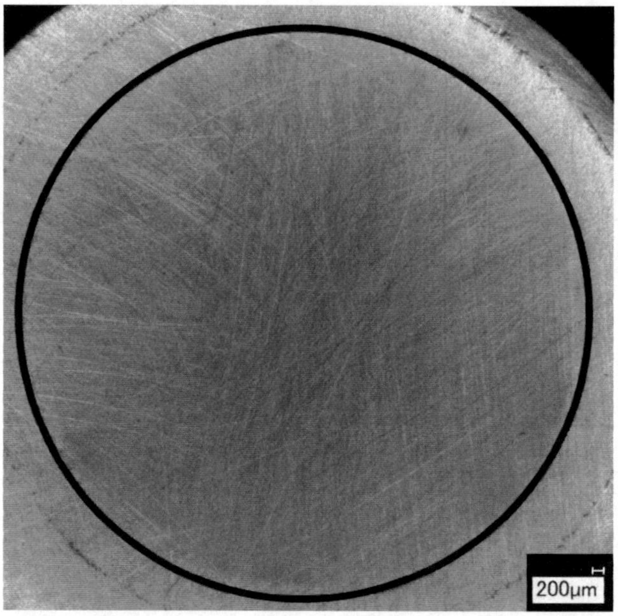

Figure 10: Visual appearance of the sample surface after the prolonged test with a temperature cycle (Figure 9); the active sample surface is within the black circle and shows a slight brownish discoloration and few local corrosion spots, but no signs of heavy corrosion. Therefore, it is plausible that corrosion initiation took place in the 6th temperature cycle, after which the test was cancelled. The contact area of the ring seal lies on the outside the black circle.

Conclusion

A new phenomenological approach for the testing and quantitative evaluation of VCI-containing products has been presented. By cooling of a metallic sample in a test chamber, in which the sample is placed in a sample holder alongside a VCI-containing product, a water film is formed by condensation of humidity from the test chamber atmosphere. The water film wets the sample surface and the auxiliary electrodes and thus allows for electrochemical measurements (EIS, DC polarization).

The obtained values of the electrochemical parameters have been correlated to the surface state of the sample with regard to the occurrence of corrosion. The pseudo-capacitance representing the capacitive behavior of the interface between sample and water film as well as the polarization resistance representing the charge transfer rate within the system are suitable parameters for the objective assessment of the surface state of the sample during climatic stress. The response of the parameters to variations of the test (amount of VCI, conditioning time) has been proven as plausible. Statistical analysis of the obtained results has shown that the test is capable of discriminating between different levels of corrosion-inhibiting effects. Further insight into the corrosion behavior of a sample under the condensed water film has been gained by time-resolved registration of the electrochemical parameters. In summary, the new test allows for a quantitative assessment of VCI-containing products on a phenomenological basis with the mode of action and the environmental conditions, which represent the corrosive stress, being similar to those during service of the VCI-containing product.

Acknowledgement

The authors gratefully acknowledge the financial support of the project by the German Federal Ministry for Economic Affairs and Energy within the framework of the ZIM funding program.

References

1. A. Subramanian et al., Corrosion, **56** (2000) 144.
2. S. Koehler and G. Reinhard, International Journal of Corrosion and Scale Inhibition, **3** (2014) 286.
3. Y. I. Kuznetsov, International Journal of Corrosion and Scale Inhibition, **4** (2015) 15.
4. N. N. Andreev and Y. I. Kuznetsov, International Journal of Corrosion and Scale Inhibition, **1** (2012) 16.
5. N. N. Andreev and Y. I. Kuznetsov, International Journal of Corrosion and Scale Inhibition, **1** (2012) 146.
6. J. E. B. Randles, Discussions of the Faraday Society, **1** (1947) 11.
7. K. Kim et al., Advances in Materials Science and Engineering, **2015** (2015) 408138.
8. Y. J. Tan, S. Bailey and B. Kinsella, Corrosion Science **38** (1996) 1545.
9. K. F. Khaled and S. S. Abdel-Rehim, Arabian Journal of Chemistry **4** (2011) 397.
10. A. Popova et al., Corrosion Science **45** (2003) 33.
11. J. W. Eaton, D. Bateman, S. Hauberg and R. Wehbring, GNU Octave version 4.2.0 manual: a high-level interactive language for numerical computations. URL http://www.gnu.org/software/octave/doc/interpreter
12. P. Agarwal, M. E. Orazem and L. H. Garcia-Rubio, Journal of the Electrochemical Society **142** (1995) 4159.

Corrosion behaviour of Fe and Ni based alloys in the reactor atmosphere of a thermal cracking process

<u>Alexander Schmid</u>[1], Gregor Mori[1], Stefan Hönig[2], Roland Haubner[3]

1 Chair of General and Analytical Chemistry, *Montanuniversität Leoben, Franz Josef Straße 18, A-8700 Leoben, Austria*
2 OMV E&P GmbH, Techcenter & Lab, Proteserstraße 40, A2230 Gänserndorf
3 Technische Universität Wien, Institute of Chemical Technologies and Analytics, Getreidemarkt 9/164, A-1060 Vienna, Austria

The alloys 1.7386, 1.4841, 1.4959 and 2.4816 were tested under conditions simulating a thermal cracking process. Tests were performed at 480, 580 and 680 °C, for 24, 72 and 240 h in a gas atmosphere containing 3.8 vol.% HCl, 200 ppm H_2S and CO, CO_2, H_2 and N_2. It represents the partial pressures, which are present in the reactor of a thermal cracking process of anthropogenic resources.
Flat samples were prepared from pipes and in order to achieve a uniform surface the samples were ground to 1000 grit SiC paper. After corrosion experiments the samples were analyzed by metallography, SEM/EDX and XRD. Additionally the mass loss of the samples during the experiment was evaluated.
During corrosion the metallic phase is attacked by HCl and various corrosion products are formed by reaction with CO_2 and H_2S from the gas phase. After the tests the corroded samples showed a porous multilayered surface structure. The outer layer consisted mainly of Cr sulphide, followed by a Cr oxide layer. Below these two layers an area containing chlorine was detected. The porosity in the corrosion zone can be explained by the evaporation of $FeCl_2$ which was observed as colorless crystals at the colder parts of the testing equipment. This general structure of corrosion products was detected at all tested materials.
At high testing temperatures the mass loss decreased with increasing Ni content. However, at lower testing temperatures this effect changed entirely, so that at 480 °C the Ni content showed no influences and the mass losses were nearly equal.

<u>Keywords:</u> "high temperature corrosion" "HCl" "H_2S"

Introduction

Due to their convenient properties, such as low price, low density and high durability, plastics are one of the most important materials in our society. However, compared to other materials the lifetime of plastic materials, especially packing plastics is very short. In 2016 16.7 million tons of post-consumer packaging plastics ended up in waste streams in the European Union. 27.3 % of them are still landfilled. Due to its low density, large volumes of plastic waste are produced, which therefore need to be disposed of [1]-[4].

The thermal cracking of post-consumer plastics to produce short chained hydrocarbons is a promising alternative. Nevertheless, halogenated plastics and biological impurities lead to HCl and H_2S during thermal cracking. Especially HCl is known as very corrosive to metallic materials above 450 °C [4]-[6].

The general accepted mechanism of chlorine induced high temperature corrosion is:

- Migration of HCl or Cl_2 through the initial oxide layer of the steel
- Formation of volatile metal chlorides below the initial oxide layer
- Evaporation of metal chlorides leading to disruption of the oxide layer
- Metal chlorides form metal oxides and HCl or Cl_2 in contact with H_2O or oxygen in the gas atmosphere
- HCl or Cl_2 can reenter the reaction cycle and act autocatalytic
- The formation of voluminous and non-protective oxide layers lead to continuous metal loss due to chloridation and subsequent oxidation of the chlorides

Therefore, the mechanism is called active oxidation, which was investigated by various authors and is well known [7]-[15]. Due to the high volatility of Fe chlorides, Fe is known as a detrimental alloying element in HCl containing environments. Ni on the other hand was found to be beneficial because of the much lower volatility of $NiCl_2$ [16]-[19].

The addition of H_2S to HCl containing gas atmospheres at high temperatures was only investigated by a few authors. No influence on the corrosion mechanism was proposed. Only higher mass losses have been reported above 400 °C, compared to gas atmospheres containing only HCl and no H_2S [20]-[22].

The aim of this work was to find a durable reactor material for a thermal cracking process of post-consumer plastics. In order to get an overview of the corrosion behaviour at the thermal cracking conditions, experiments were performed with construction materials with a wide range of Fe and Ni contents, starting from 1.7386 containing no Ni till 2.4816 with 72.5 % Ni.

Experimental

The detailed experimental procedure and the metallographic preparations can be found elsewhere [23],[24]; the chemical composition of the tested materials is shown in Table 1.

All specimens were polished with 1000 grit SiC paper, washed, degreased and gauged, in order to reach a well-defined substrate surface. The gas for the tests represents the corrosive gas atmosphere in a thermal cracking process for post-consumer plastics and consisted of 3.8 vol.% HCl, 1.9 vol.% CO_2, 0.3 vol.% CO, 2.8 vol.% H_2, 0.02 vol.% H_2S, bal. N_2. It was provided in two different bottles as HCl cannot be stored together with H_2S. The corrosion experiments were performed between 480-680 °C in a silica glass tube for ten days. The selection of the testing temperatures was based on the knowledge gained from a small pilot plant.

Table 1: Alloying composition of the tested materials

Material	Fe	Ni	Cr	C	Mn	Si	Ti	Al	Mo
1.7386	89		9	0.1	0.5	0.5			1
1.4841	52	20	25	0.07	1.5	1.5			
1.4959	47.5	30.4	20.5	0.07	0.6		0.5	0.5	
2.4816	<10	72.5	16.5	<0.05	0.8	0.4			1

Results and Discussion

The visual appearance of the samples after the corrosion experiments is depicted in Figure 1. All tested alloys formed loose, black corrosion products at 680 °C. At lower testing temperatures these black layers were also observed on materials with lower Ni content. Only 2.4816 with 72.5 % Ni formed silver shiny layers, which were also more adherent, at 480 and 580 °C.

After all experiments colorless crystals were observed outside the heated zone in the colder parts of the silica tube. Their amount varied depending on the testing temperature and the tested material. More of these crystals were found at higher temperatures and after tests with materials with higher Fe content. These colorless crystals were determined to be pure $FeCl_2$ with EDX and XRD analyses. It evaporated from the samples, was carried away with the gas flow, and crystallized at colder parts of the silica tube. Figure 2 shows the powder diffractogram of the colorless crystals.

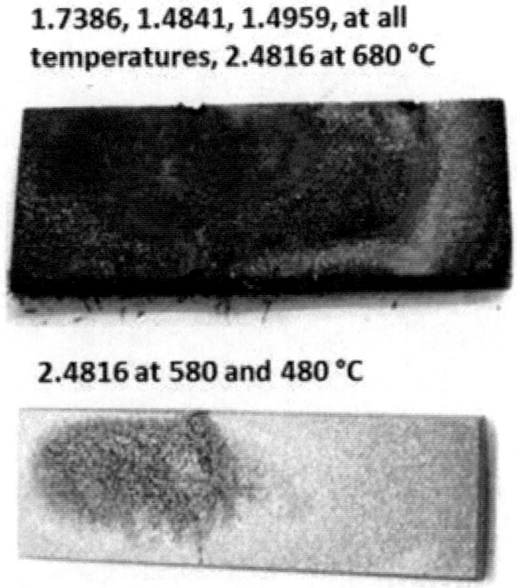

Figure 1: Different materials after a corrosion experiment

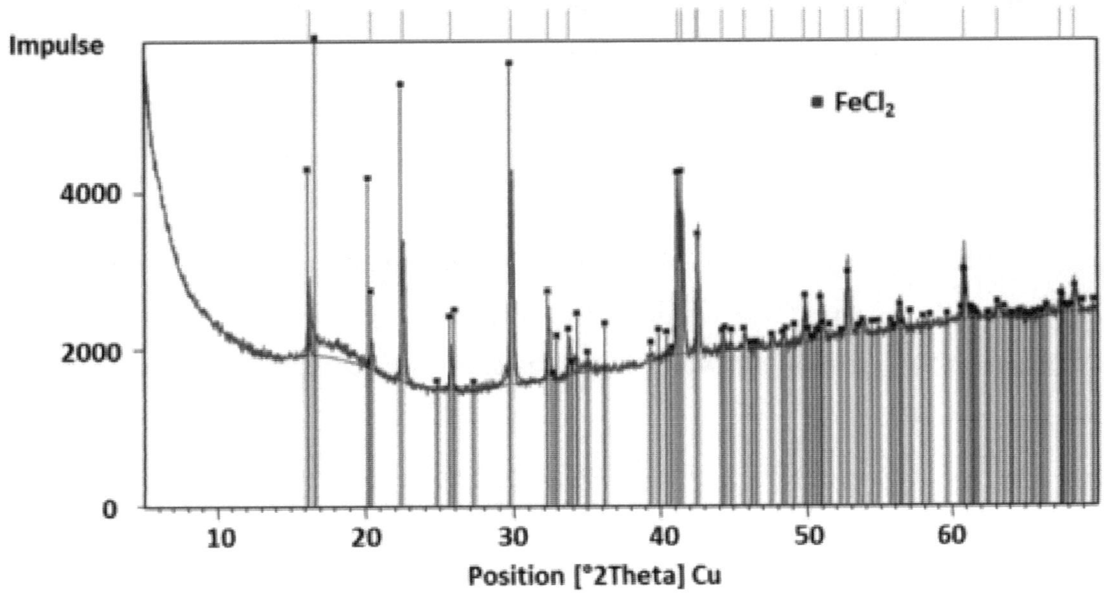

Figure 2: Powder diffractogram of the colorless crystals, observed in the cold parts of the testing equipment

Corrosion rates for the different steels at various corrosion conditions

Figure 3 displays the mass losses of the tested materials after one, three and ten days at 680 °C. There is a correlation between the Ni content of the alloys and their mass loss during the corrosion experiments. 1.7386 contains no Ni and showed the highest mass loss after one as well as after three days at 680 °C. As the sample was almost consumed after the three day experiment, no ten day experiment was carried out. After three days experiment the material had a mass loss of 234 mg/cm². Since no ten days experiment could be performed for 1.7386, the results do not allow a conclusion of the mass loss evolution of this material. However, literature suggests linear mass loss rates for low alloyed steels in chlorine containing environments above 500 °C [25].

1.4841 with 20 % Ni showed a comparably low mass loss of 35 mg/cm² after ten days. 1.4959 contains about 30 % Ni and 2.4816 with 72.5 % Ni showed even lowest mass losses of 5 respectively 3 mg/cm² after ten days at 680 °C. Moreover, the mass loss of 1.4959 and 2.4816 did not increase significantly between three and ten days. Also the mass loss rates of 1.4841 decreased with time, but not as significant as for the two higher Ni containing alloys.

The decreasing mass loss rates with testing time reveal the formation of corrosion products, which retard the corrosion process. As Figure 4 illustrates, at 580 °C 1.7386 showed a much lower mass loss compared to 680 °C (40 mg/cm² after ten days). Also the mass loss did not change significantly between three and ten days. The correlation between the decreasing mass losses and the increasing Ni content of the alloys is also present, such as at 680 °C. Only the difference between the materials is not as significant.

While the mass loss of 1.4841 decreased from 35 to 12 mg/cm² after ten days at 580 °C, the mass loss of 1.4959 did not change with the 100 °C temperature decrease. 2.4816 even increased its mass loss to 4 mg/cm² at 580 °C after ten days.

At 480 °C all materials showed very similar mass losses (Figure 5). While the mass losses of the low Ni steels decreased with temperature as expected, the mass losses of high Ni steels (1.4959 and 2.4816) increased to around 10 mg/cm² after ten days (Figure 6). It shows coherence between the Ni content of an alloy and the influence of the temperature on its mass loss. Low Ni containing alloys decreased their mass loss with decreasing temperature, while materials with high Ni content increased their mass loss with decreasing temperatures. As corrosion rates generally decrease with temperature because of slower reaction kinetics, the results of the two high Ni containing materials were unexpected.

In order to be able to explain these findings, cross sections of corrosion products were prepared and analyzed via SEM/EDX and XRD. Figure 7 depicts a powder diffractogram of the corrosion products on the sample at 680 °C. Sulphides and the oxide of Cr were detected.

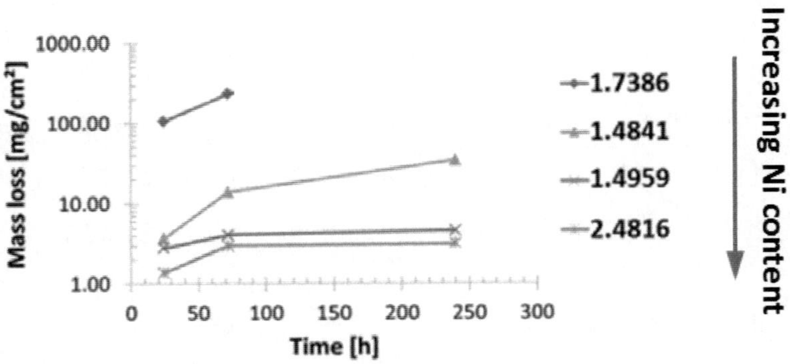

Figure 3: Mass loss vs. time of the tested materials, 680 °C (value for 1.7386, 10 d could not be measured as the sample was almost consumed after 3 d)

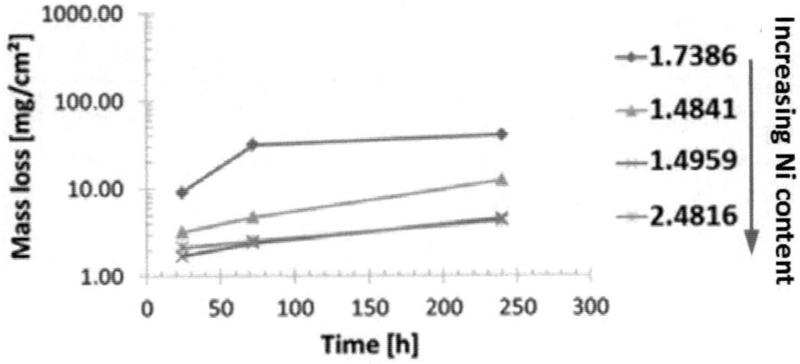

Figure 4: Mass loss vs. time of the tested materials, 580 °C

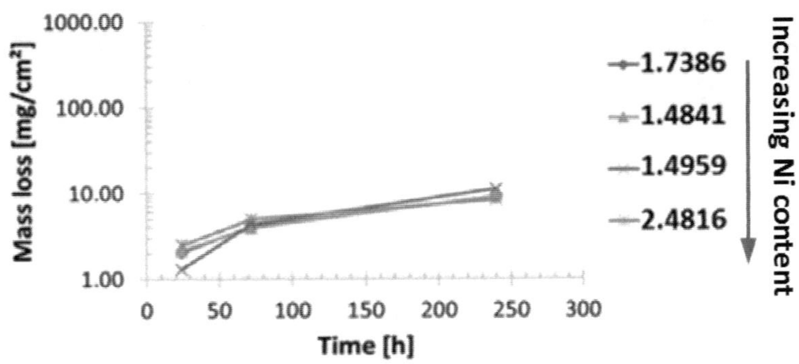

Figure 5: Mass loss vs. time of the tested materials, 480 °C

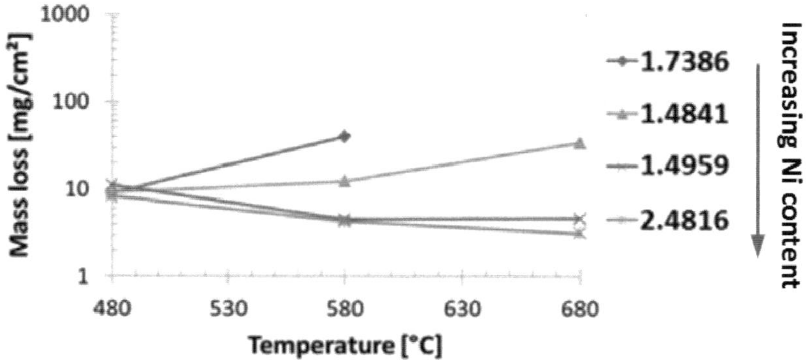

Figure 6: Mass loss vs. temperature of the tested materials after 10 d corrosion.

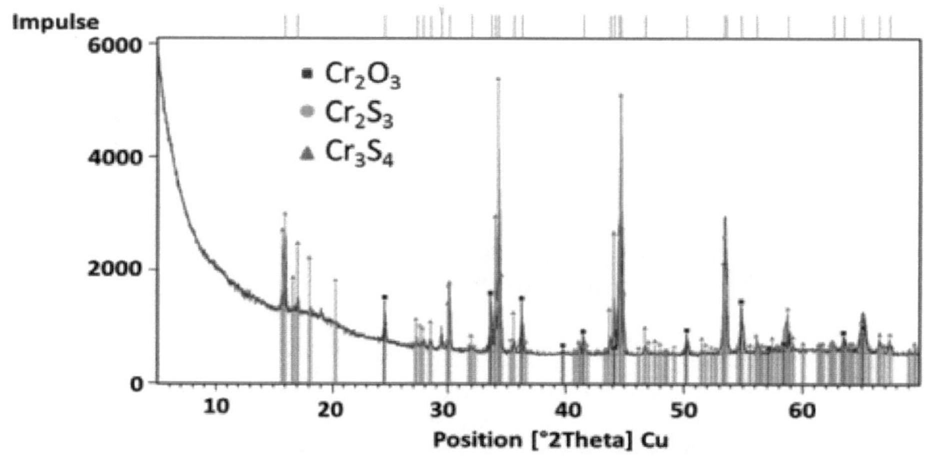

Figure 7: Powder diffractogram of corrosion products detected on the sample

Microstructure of the observed corrosion layers

Figures 8-10 illustrate the SEM images of cross sections and the corresponding EDX mappings of 2.4816 at different temperatures. This material was chosen exemplary, as the general structure of corrosion products did only change with temperature, but was identical on all tested materials. An EDX mapping of 1.4841 can be found in [23].

At 680 °C sulphides of Cr were detected above a layer containing Cr and oxygen (Figure 8). This is in good agreement with the XRD analysis, which detected Cr_2S_3, Cr_3S_4 and Cr_2O_3. Chlorine was observed in between the small pores below the oxide layer. Further, an enrichment of Ni and a depletion of Cr compared to the metal matrix were detected in the border area next to the corrosion products. Also Fe depleted in this area. But as 2.4816 contains only < 10 % Fe, this effect is hardly visible in the mapping. The depletion of Fe and Cr can be explained by the formation of the observed corrosion products. Ni was not detected in any corrosion product. Thus it remained in the metal and showed no chemical reaction with the gas atmosphere.

At 580 °C the EDX mappings reveal the formation of Ni sulphides (Figure 9). Besides also sulphides of Cr and Cr_2O_3 were observed in the corrosion products. Such as at 680 °C the sulphides formed above the oxide. A chlorine containing layer was detected below the oxide. Compared to 680 °C the corrosion scales on 2.4816 were more voluminous at 580 °C.

As Figure 10 displays, at 480 °C the corrosion scales on 2.4816 were even thicker than at 580 °C. Instead of big Cr sulphide crystals, big Ni sulphide crystals were observed above a Cr oxide layer. Also small amounts of Fe were found in the Ni sulphide crystals. Chlorine was still detected below the oxide. The amount of detected chlorine increased with decreasing temperature. However, it is unlikely that more metal chlorides formed at lower temperatures than at higher ones. More probably the lower volatility of the metal chlorides at lower temperatures is responsible for the findings. At 680 °C much more metal chlorides evaporated from the surface, while at 480 °C they remained below the corrosion products.

As mentioned above all other materials showed the same structure of corrosion products at the respective temperatures. Only the amount of the sulphides and oxides changed with the material. 1.7386 formed no Ni containing corrosion products as this alloy involves no Ni. This is also in good agreement with the strongly decreasing mass losses of this material with decreasing temperature, as the EDX mappings reveal the formation of Ni sulphides to be responsible for the high mass losses at lower temperatures. In opposite to 2.4816 and 1.4959, the lower Ni containing alloys 1.7386 and 1.4841 formed thinner corrosion layers with decreasing temperature.

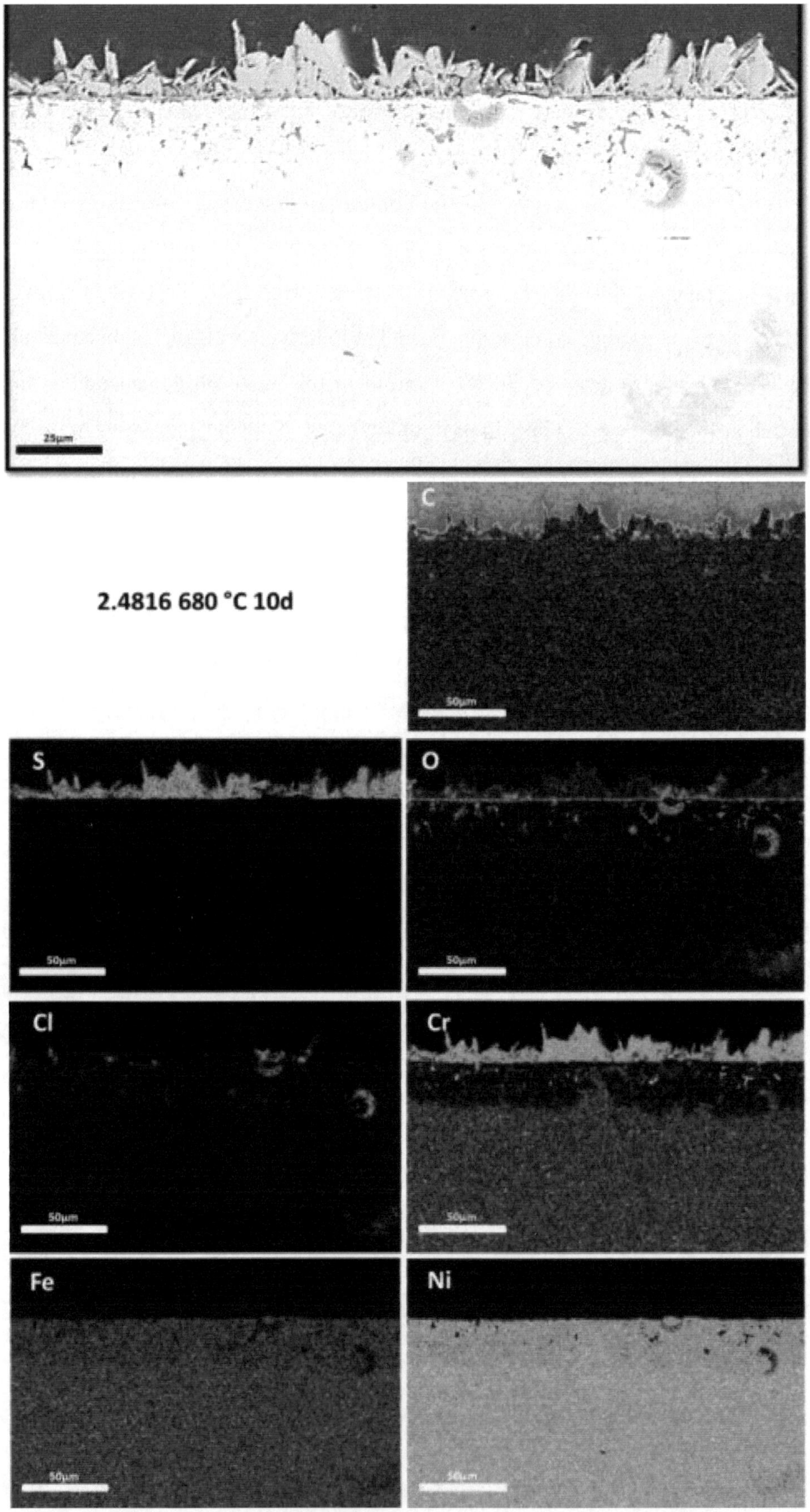

Figure 8: EDX mapping of 2.4816, after a ten day experiment, at 680 °C

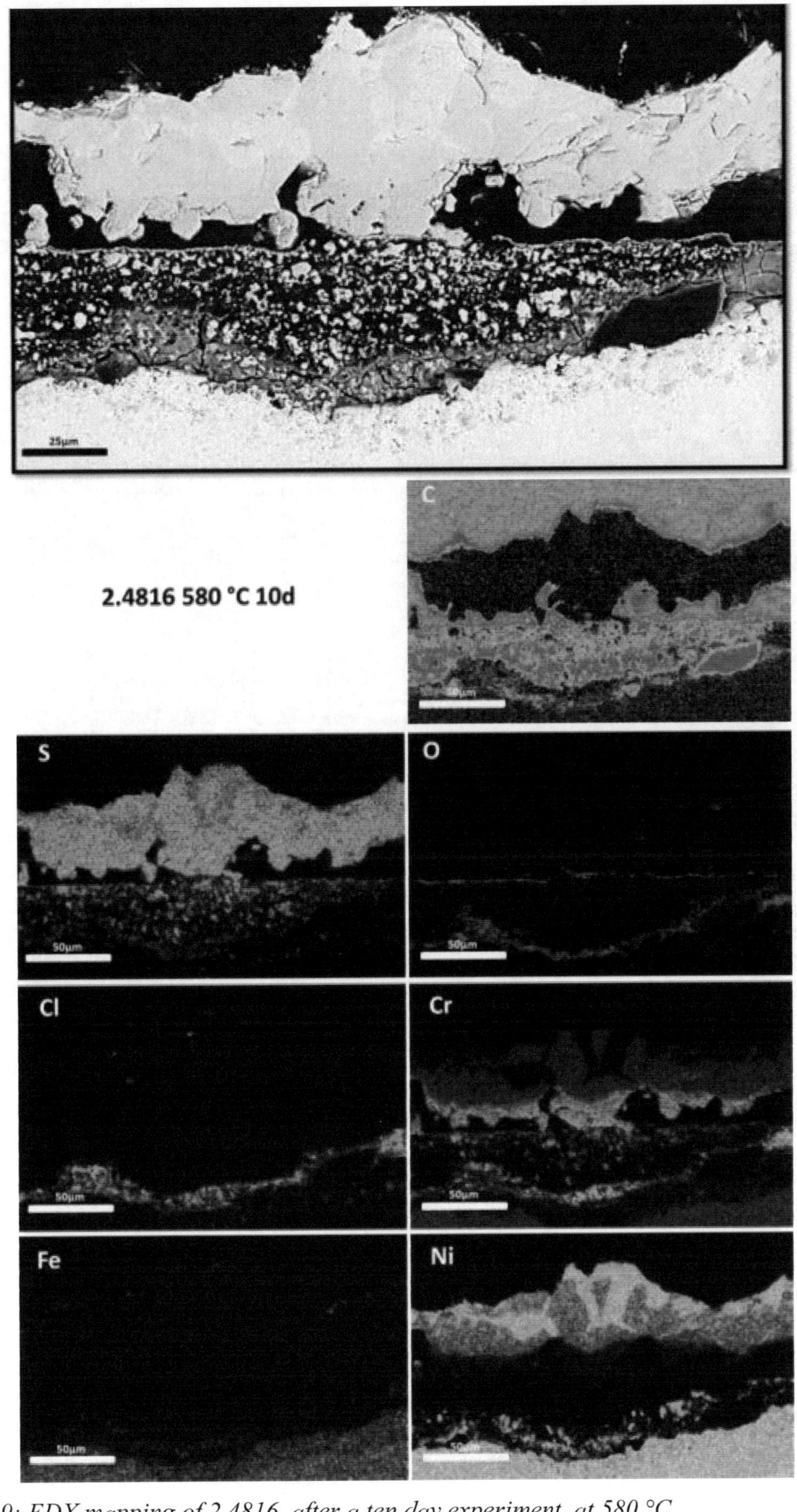

Figure 9: EDX mapping of 2.4816, after a ten day experiment, at 580 °C

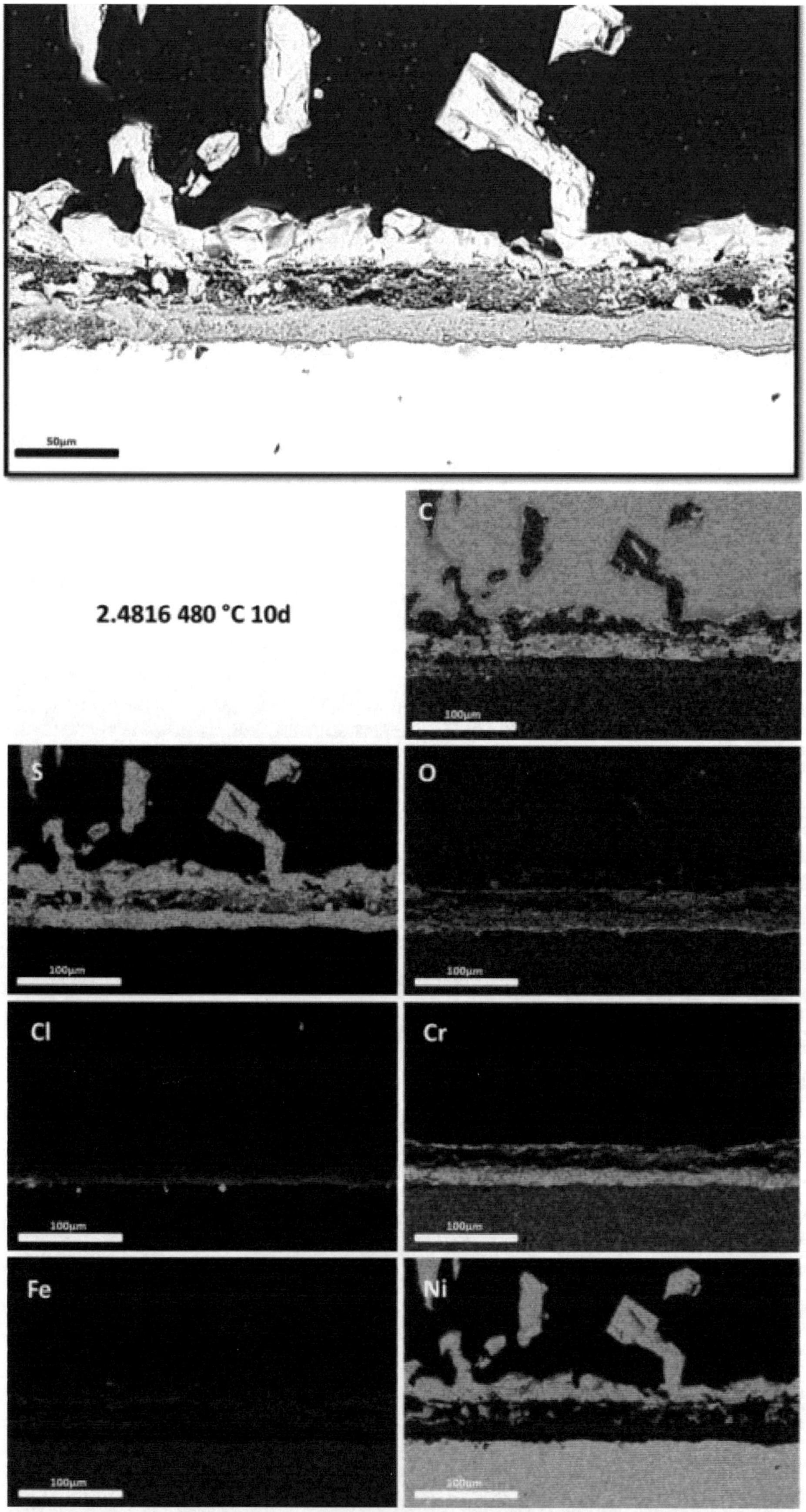

Figure 10: EDX mapping of 2.4816, after a ten day experiment, at 480 °C

Explanation for the formation of the layered structure of corrosion products

In the following an explanation for the formation of the corrosion products and their particular structures at the respective temperatures should be found. All tested materials contain an amount of Cr, which is sufficient to form a Cr oxide layer on the metal samples. Thus, they are protected against the carbon containing gases and H_2S. HCl is able to move through the initial oxide scale, to form metal chlorides below it via the following reaction.

$$Me + 2\ HCl \rightarrow MeCl_2 + H_2 \qquad Me = Fe, Cr, Ni$$

Calculations with the software FactSage 7.1 ® showed that this reaction only takes place for Fe and Cr, as the Gibbs free energy of formation of $NiCl_2$ via this reaction is positive. This explains the beneficial influence of Ni on chlorine induced high temperature corrosion. $FeCl_2$ is volatile at all tested temperatures and evaporated constantly from the sample. It was detected as pure $FeCl_2$ in the colder parts of the silica tube behind the furnace. $CrCl_2$ has a much lower vapor pressure than $FeCl_2$ and evaporates slowly at 680 °C. At 480 °C the evaporation of $CrCl_2$ is negligible.

The detected formation of Cr_2O_3 scales requires an oxygen donator. No molecular oxygen is involved in the gas phase and CO_2 is not sufficient as the Gibbs free energy of the reaction of CO_2 and $CrCl_2$ to Cr_2O_3 is highly positive. But, CO_2 can react with H_2 via the watergas-shift reaction to form CO and H_2O. This formed water can react with the evaporating $CrCl_2$ to form Cr_2O_3 and HCl. HCl is able to reenter the reaction cycle and act autocatalytic. Since the watergas-shift reaction is catalyzed by metallic and oxide surfaces, this reaction will take place on the sample surface and after the corrosion reaction progress also in the corrosion scales.

The equilibrium of the watergas-shift reaction is shifted to the educts side at lower temperatures. Thus, at 480 °C less H_2O will be produced than at 680 °C. Consequently, more Cr_2O_3 is formed at 680 °C, resulting in a denser layer than at 480 °C. This is especially true for 1.4841, which contains 25 % Cr [24].

The reason for the big Cr sulphide crystals, which formed at 680 °C, can be found in the reaction of $CrCl_2$ with H_2S. Excessive $CrCl_2$, which had not reacted with H_2O, reacts with H_2S to form the detected phases Cr_2S_3 and Cr_3S_4. Since the oxide scale is denser at 680 °C and more $CrCl_2$ evaporate compared to 480 and 580 °C, H_2S cannot reach deeper regions of corrosion products. Thus, the sulphides are formed on top of the corrosion products and metallic Ni is protected from H_2S.

At lower temperatures H_2S is able to reach the metallic surface, where it reacts with Ni to form voluminous sulphides. Since at 580 °C still small amounts of $CrCl_2$ evaporate, mixed Cr and Ni sulphide crystals formed (Figure 9). At 480 °C only the sulphides of Ni formed the voluminous crystals (Figure 10).

Recapitulation of the observed results

Summarizing the corrosion behavior and the mass losses of the different tested materials can be explained. 1.7386 contains 91 % Fe which evaporates as very fast $FeCl_2$ at 680 °C. The formed scales of Cr_2O_3 and Cr sulphides had no protective effect, resulting in the very high mass loss compared to the other tested materials. At lower testing temperatures, smaller amounts of $FeCl_2$ and $CrCl_2$ evaporated, which also results in lower mass losses of 1.7386.

On the other hand 2.4816 contains < 10 % Fe and >72 % Ni. At 680 °C also large amounts of $FeCl_2$ and $CrCl_2$ evaporate from the sample, resulting in the corrosion scales. But as these two elements are consumed in the border areas near the surface and Ni remains, HCl is not able to move to lower regions of the sample. Thus, the remaining Ni is a sufficient diffusion barrier for HCl and the corrosion process stops as long as it is protected from H_2S. This explains the low mass loss of 2.4816 at 680 °C. At lower temperatures Ni is consumed due to the reaction with H_2S, resulting in higher mass losses.

1.4841 and 1.4959, with amounts of alloying elements between 1.7386 and 2.4816 fit into the scheme and also showed mass losses between those two extreme cases.

Conclusion

The alloys 1.7386, 1.4841, 1.4959 and 2.4816 were investigated in an atmosphere simulating thermal cracking conditions of post-consumer plastics. Chlorine induced high temperature corrosion was found to be the dominating corrosion mechanism, which was influenced by the volatility of the evaporating metal chlorides. While the general corrosion mechanism did not change with temperature, it had a big influence on the mass loss of the different tested materials.

Generally high Ni contents in steels are beneficial at higher temperatures of corrosion (see test series at 680 °C). At lower temperatures Ni reacts easier with H_2S to form Ni sulphides and corrosion rates increases. Therefore, it is not possible to define one material, which will perform best at all conditions, but to define an operating window for the different materials.

Below 500 °C 1.7386 showed sufficiently low mass losses. On the other hand the Ni alloys require operation temperatures above 600 °C.

Acknowledgement

Mr Matthias Weil from the Technische Universität Wien, performed the XRD analysis which is gratefully acknowledged. Thanks are also due to Sandvik who provided the materials for the tests.

Literature

[1] Plastics – the facts 2016: An analysis of the European plastics production, demand and waste date, **2017**

[2] Plastics Europe. Plastics – the facts 2017: An analysis of the European plastics production, demand and waste date, **2018**

[3] José Aguado, David P. Serrano, Feedstock recycling of plastics waste, RSC clean technology monographs, Royal Sec. of Chemistry, Cambridge, **1999**

[4] European Commission, Proposal for a Directive of the European Parliament and of the Council amending Directive 94/62/EC on packaging and packaging waste, **2015**/0276 (COD)

[5] J. Yu, L. Sun, C. Ma, Y. Qiao, H. Yao, Thermal degradation of PVC: A review, Waste Management, **2015**

[6] G. Sorell, The role of chlorine in high temperature corrosion in waste-to-energy plants, G. Sorell Consulting Services, Materials at high temperatures 14, **1997,** 137-150

[7] A. Zahs, M. Spiegel, H.J. Grabke, The influence of alloying elements in the chlorine induced high temperature corrosion if Fe-Cr alloys in oxidizing atmospheres, Materials and Corrosion, **1999**, 50, 561-578

[8] H. Asteman, M. Spiegel, Investigation of the HCl (g) attack on pre-oxidized pure Fe, Cr, Ni and commercial 304 steel at 400 °C, Corrosion Science, **2007**, 49, 3626-3637

[9] D. Bramhoff, H.J. Grabke, H.P. Schmidt: in: The role of active elements, Ed. E. Lang, Elsevier Appl. Sc., **1989**, 335-349

[10] W.T. Bakker, The effect of chlorine on mixed oxidant corrosion of stainless steel, Materials at high temperatures, 1997, 14, 197-206

[11] S.H. Lee, M.J. Castaldi, The effects of varied hydrogen chloride gas concentrations on corrosion rates of commercial tube alloys under simulated environment of WTE facilities, Waste-to-Energy conference, Pennsylvania USA, 2008

[12] R. Prescott, F.H. Stott, P. Elliott, The degradation of metals in a hydrogen chloride containing gas at high temperature, Corrosion Science, **1989**, 29, 465-475

[13] D. Bramhoff, H.J. Grabke, H.P. Schmidt, Influence of HCl and N_2 on high temperature corrosion in atmospheres with low oxygen pressures, Werkstoffe und Korrosion, **1989**, 40, 642-650

[14] V.A.C. Haanappel, T. Fransen, P.J. Gellings, Chlorine-induced high temperature corrosion: I. Metals and Alloys – A review, High Temperature Materials and Processes, **1992**, 10, 67-89

[15] D. Bramhoff, H.J. Grabke, E. Reese, Influence of HCl and Cl_2 on high temperature corrosion of 2 ¼ Cr1Mo steel in atmospheres with high oxygen pressures, Werkstoffe und Korrosion, **1990**, 41, 303-307

[16] Y. Ihara, H. Ohgame, K. Sakiyama, The corrosion behaviour of iron in hydrogen chloride gas and gas mixtures of hydrogen chloride and oxygen at high temperatures, Corrosion Science, **1981**, 21, 805-817

[17] Y. Ihara, H. Ohgame, K. Sakiyama, The corrosion behaviour of chromium in hydrogen chloride gas and gas mixtures of hydrogen chloride and oxygen at high temperatures, Corrosion Science **1983**, 23, 167-181

[18] Y. Ihara, H. Ohgame, K. Sakiyama, The corrosion behaviour of nickel in hydrogen chloride gas and gas mixtures of hydrogen chloride and oxygen at high temperatures, Corrosion Science, **1982**, 22, 901-912

[19] A. Schmid, G. Mori, S. Strobl, R. Haubner, S. Hönig: "Corrosion of various Fe and Ni based alloys in HCl, H_2S containing environments, with low oxygen partial pressure, at 680 °C"; Presentation: EUROCORR 2017, 20th International Corrosion Congress & Process Safety Congress 2017, Prague, Czech Republic; 03.09. - 07.09.**2017**; in: "*Conference Proceedings EUROCORR 2017*", (2017), Paper-Nr. 79016, 10 S.

[20] W.T. Bakker, R.A. Perkins, Beyond mixed oxidant corrosion-corrosion Phenomena in gasifiers, EPRI Gasification Conference, **1990**

[21] W.T. Bakker, R.A. Perkins, in: Materials for coal gasification, ASM International, W.T. Bakker, S. Dapkunas and V. Hill (Eds.), **1988**, 85-96

[22] T.J. Pan et.al., Corrosion behaviour of three iron-based alloy in reducing atmospheres containing HCl and H_2S at 600 °C, Corrosion Science, **2007**, 49, 1362-1377

[23] A. Schmid, G. Mori, R. Haubner, M. Weil, S. Hönig, Materials and Corrosion, **2018**, 1

[24] A. Schmid, G. Mori, S. Hönig, M. Weil, S. Strobl, R. Haubner, Comparison of the high-temperature chloride-induced corrosion between Duplex steel an Ni based alloy in presence of H_2S, Corrosion Science, **2018**

[25] N.S. Jacobson, Reaction of Iron with Hydrogen Chloride-Oxygen Mixtures at 550 °C, Oxidation of Metals, **1985**, 26, 157-169

Influence of concentration and coating time on the barrier formation of sustainable N-acyl sarcosines to protect low carbon steel in aqueous solution.

Saad KASKAH[1], Gitta EHRENHAFT[2], Jörg GOLLNICK[2], Christian FISCHER[3]

[1] *University of Koblenz-Landau, Germany*

[2] *Institute of mechanics and material science, TH Mittelhessen University of Applied Sciences, Germany*

[3] *Department of Physics, University Koblenz-Landau, Germany*

Abstract

Three N-acyl sarcosine derivatives were tested for their potential as inhibitive layers forming on low carbon steel CR4 against corrosion in 0.1 M NaCl and normal water. The protective barrier is achieved by direct adsorption on the steel surface via dip coating in compound stock solutions. Variations in sarcosine concentration and coating time were applied for different evaluation systems (weight loss, electrochemical, and spray corrosion measurements) to identify best conditions.

A highly improved corrosion resistance for steel CR4 with a direct correlation to the molecular weight resp. carbon chain length of present N-acyl sarcosines compounds was revealed. Efficiencies enhanced for each substance to a certain degree with rising concentration and extended dip coating time. Oleoylsarcosine (O) and Myristoylsarcosine (M) were best to reduce corrosion at the highest investigated concentration of 100 mmol/L with efficiencies up to 97.2 % and 82.5 % in the polarization tests. 51.2 % was the result for Lauroylsarcosine (L). The best exposure time for the CR4 dip coating turned out to be 10 min for all tested compounds. The adsorption process of tested compounds is assumed to follow the Langmuir type adsorption.

For a more realistic approach, endurance tests were performed 24 hours in a spray corrosion chamber with normal water. Each sample was digitalized with a scanner after specific times and the corroded area analyzed with the image-processing program ImageJ. The obtained trend to decrease corrosion of CR4 steel was confirmed by efficiencies for L with 71.4 %, M with 79 % and O with 99.6 %.

The effect of electrolyte flow rate on corrosion inhibition by self-assembled films of long-chain phosphonic acid

Ekatarina KRISTAN MIOČ[1], Benjamin RIAUTET[2], Zana HAJDARI GRETIĆ[3], Helena OTMAČIĆ ĆURKOVIĆ[1]

[1]*Faculty of Chemical Engineering and Technology, University of Zagreb, Zagreb, Croatia,* helena.otmacic@fkit.hr
[2]*National Polytechnic Institute of Chemical Engineering and Technology, Toulouse, France,* benjamin.riautet@ensiacet.fr
[3]*Pliva Croatia d.o.o., Zagreb, Croatia,* zana.hajdarigretic@pliva.com

Abstract
Phosphonic acids are known for their spontaneous self-assembly on oxidized substrates. They easily chemisorb on many metals and alloys and show good resistance to hydrolysis and therefore they can presents potential ecologically suitable solution for increasing a corrosion resistance of alloys under corrosive condition. While phosphonic acids have become popular surface modifiers for metal oxides in various fields, especially in the past decade, there is still a lack of research on their stability under realistic application condition. One of the parameters influencing the corrosion of equipment materials is the flow velocity, through its effect on mass transfer and other phenomena such as erosion-corrosion. Therefore, the aim of this work is to study the effect of the electrolyte flow rate on corrosion of, unprotected and phosphonic acid modified, cupronickel and stainless steel alloys in river and seawater. The investigation is conducted at laboratory scale by the use of flow loop to simulate real conditions. Electrochemical polarization methods and electrochemical impedance spectroscopy are used to determine protective properties of organic film and the corrosion rate of unprotected alloys as function of flow velocity.

Keywords: flow condition; long-chain phosphonic acid; self-assembled layers.

Introduction

Many industrial plants on the shore or offshore, such as plants for conditioning and desalination of water, oil, paper, food and nuclear industries and oil platforms, often use seawater or river water, for example as a coolant in heat exchangers. Although highly resistant alloys, such as cupronickel alloy or stainless steels, are used for designing these plants and equipments, they are subjected to different corrosion processes and their service life in such aggressive environment can be prolonged by protection with corrosion inhibitors. However, marine environment is sensitive to toxic compounds that are often used as corrosion inhibitors. Phosphonic acids are known for their spontaneous self-assembly on oxidized substrates. They easily chemisorb on many metals and alloys and show good resistance to hydrolysis and therefore they can presents potential ecologically suitable solution for increasing a corrosion resistance of alloys under corrosive condition [1-3]. Most of the studies on phosphonic acids SAMs are focused on their physical properties, characterization of SAM structure and initial corrosion protection [1-5], but there is still a lack of research on their stability under realistic application condition, which is very important for their practical application. Moreover, many studies have been conducted on pure metals while studies on alloys are much scarcer. One of the parameters influencing the corrosion of equipment materials is the flow velocity, through its effect on mass transfer and other phenomena such as erosion-corrosion. Above the critical flow rate, corrosion inhibitors may lose their effectiveness, which depends of the nature and molecular structure of the inhibitor as well as on its concentration and temperature. Therefore, the aim of this work is to study the effect of the electrolyte flow rate on corrosion of, unprotected and phosphonic acid modified, cupronickel and stainless steel alloys in natural seawater and river water.

Experimental

Sample preparation

Investigations were performed on cupronickel alloy rod, Cu70Ni30, and on stainless steel rod, AISI 316L. In order to prepare working electrodes, rods were cut-out in 0.5 cm thick samples and on their back side a copper wire was soldered. At the end they were embedded into epoxy resin and the exposed surface of SS and CuNi working electrode were 1.13 and 1.33 cm^2, respectively. The electrodes were abraded with emery paper grade 800, 1200 and 2500 and polished with α-Al2O3 particle size 0.1 μm, degreased with ethanol in ultrasonic bath and rinsed with deionised water. The SAM formation was carried out by the following procedure: oxide formation, 24 h at 80 °C for cupronickel and 24 h at room temperature for stainless steel, followed by spraying samples with ODPA solution, and final drying step, 5 h at 80 °C. Acid adsorption occurred in 0.01 mol dm^{-3} octadecylphosphonic acid ethanolic solution. For comparison, blank samples were prepared, on which native oxide layer was formed during 24 h at 80 °C (CuNi) or room temperature (SS). The oxidation and drying were conducted in an air convection oven.

Electrochemical measurements

Electrochemical investigations were conducted in a three electrodes cylindrical electrochemical cell, in natural seawater, sampled near the city of Zadar, and in Sava river water sampled near the city of Zagreb. Saturated calomel electrode was used as a reference electrode and graphite electrode was used as a counter electrode. All potentials in the text are referenced against saturated calomel electrode (SCE). Prior to electrochemical measurements the electrodes were immersed in the test solution for one hour to stabilize the open circuit potential. To increase the flow rate into the cell, the cell was connected to a pump and a flowmeter to fix the flow at controlled values. Experimental equipment is presented in Fig.1.

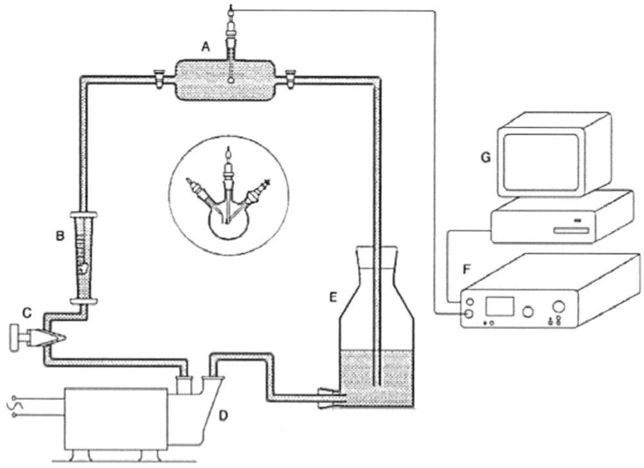

Figure 1. Equipment for electrochemical measurement under flow regime: (A) flow-through cell with three electrodes, (B) flowmeter, (C) valve, (D) pump, (E) vessel, (F) potentiostat, and (G) computer.[6]

Measurements had been performed without flow, and at three different flow rates: 0.8 L/min, 1.6 L/min and 2.4 L/min. Diameter of the cell was 4.5 cm and surface of the full section was $1.59.10^{-3}$ m^2. According to this surface and the three flow rates previously mentioned, the investigated velocities of the solution in the cell were 0.0084 m s^{-1}, 0.0168 m s^{-1} and 0.0252 m s^{-1}. The polarization measurements were performed at the narrow (± 20 mV vs. E_{oc}) potential range, with a potential scan rate of 0.166 mV s^{-1}. The electrochemical impedance spectroscopy (EIS) was performed at E_{oc} in the frequency range 100 kHz – 10 mHz with a 10 mV_{rms} amplitude. The obtained impedance spectra were interpreted on the basis of equivalent electrical circuits using ZSimpWin software. In general, $\chi 2$ value was always below $5 \cdot 10^{-4}$. The electrochemical measurements were performed using a BioLogic SP-300 potentiostat. All the measurements were conducted in triplicate.

Results and discussion

Influence of media flow rate on the stability of unprotected and ODPA modified alloys

Polarization resistance results

In order to evaluate the influence of natural media flow rate on stability of the prepared layers of octadecylphosphonic acid, untreated and treated samples were exposed to a river and seawater without flow (0 L/min) and at a flow of 0.8, 1.6 and 2.4 L/min. Polarization resistance dependence on media flow rate determined from the polarization measurements for untreated and ODPA modified cupronickel alloy is presented in Fig.1.

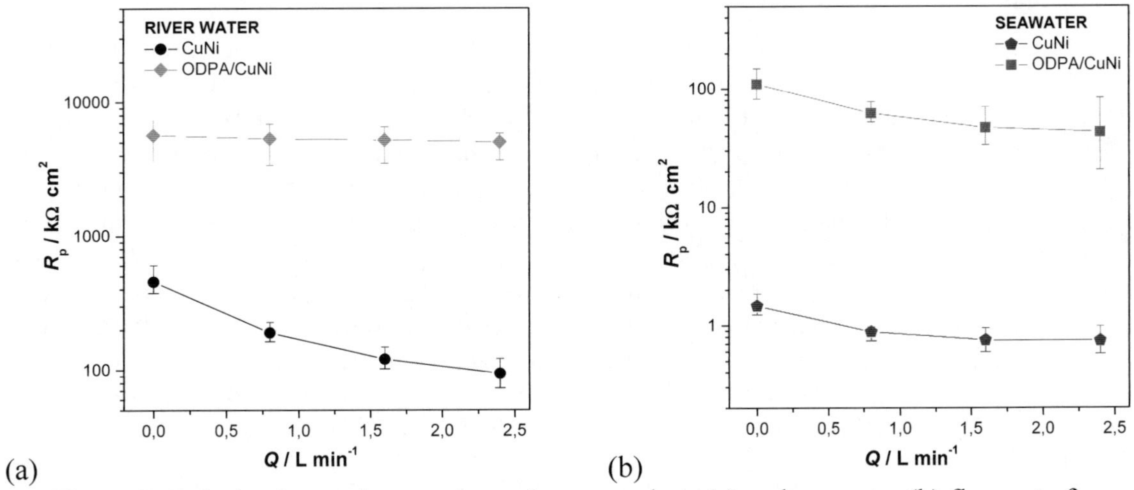

Figure 2. Polarization resistance dependence on river (a) and seawater (b) flow rate for untreated and ODPA modified cupronickel alloy.

From Fig. 2. it is evident that in both media the protective films efficiently protect copper-nickel substrate at all studied at different flow rates. For untreated samples, polarization resistance values decrease with increasing flow rates in both media which could be ascribed to more efficient O_2 transport towards the metal surface as well as to the deterioration of protective oxide film. Although the polarization resistance values of treated samples in the seawater decrease, they remain significantly higher than those for blank samples in both media, which suggests that self-assembled films are resistant to the flow of natural media at the investigated flow rates.

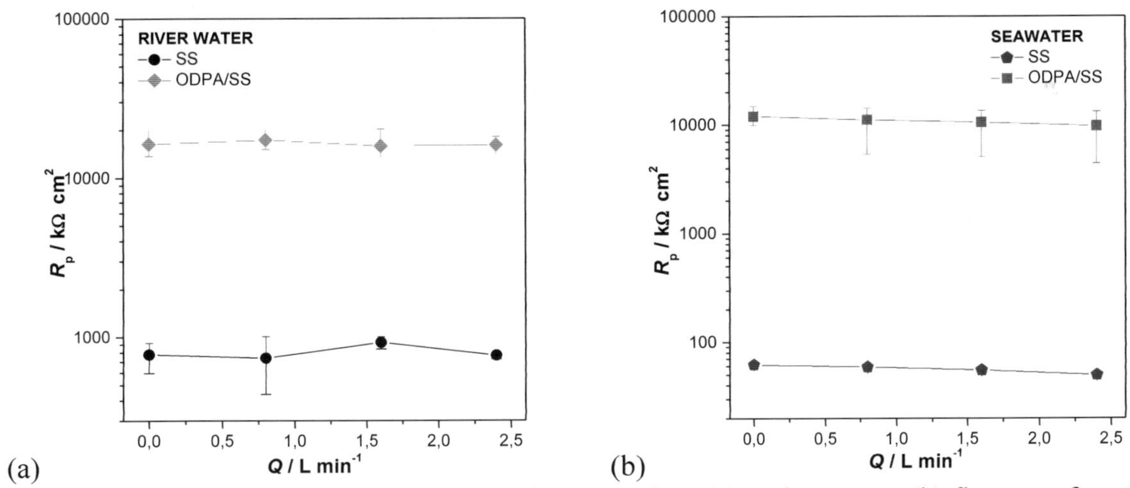

Figure 3. Polarization resistance dependence on river (a) and seawater (b) flow rate for untreated and ODPA modified stainless steel alloy.

Investigation of untreated and ODPA treated stainless steel gave similar results as for cupronickel alloys. From Fig. 3. it can be seen that in both media there is no deterioration of protective films at studied flow rates of water and the values of polarization resistance remain relatively stable and significantly higher comparing with the values of untreated samples.

Electrochemical impedance spectroscopy results

Electrochemical impedance spectroscopy measurements were conducted in order to better understand corrosion behaviour studied CuNi samples in seawater. Impedance spectra obtained for blank sample are presented in Fig. 4. It can be seen that impedance decreases with the increase of seawater flow rate. All spectra exhibit one phase maximum at medium frequencies and one at the lowest frequencies. The first one is ascribed to electrochemical reaction occurring at the metal – electrolyte interface while the second is ascribed to diffusion process. For that reason EIS spectra were modelled with the electrical equivalent circuits presented in Fig. 5 where R_{ct} represents charge transfer resistance; Q_{dl} constant phase element describing the double layer capacitance and n_{dl} corresponding coefficient. For the three lower flow rates the low frequency region was better fitted with the Warburg impedance model (W) describing semi-infinite diffusion process while for the highest flow rate it was better described with finite-diffusion element (O). The obtained impedance parameters are presented in Table 1.

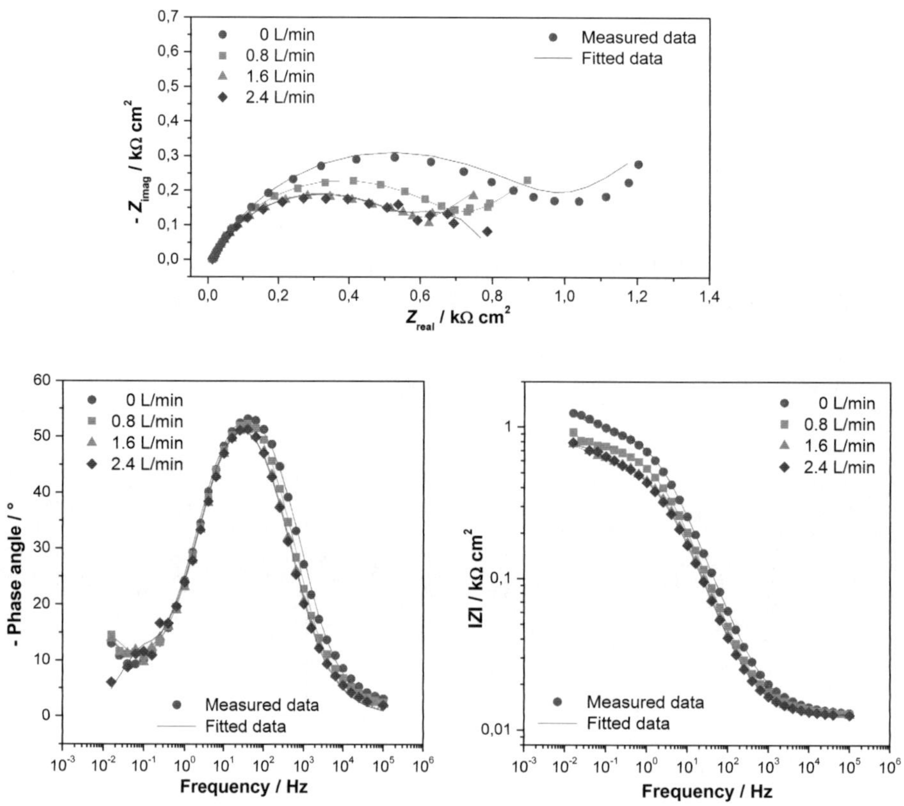

Figure. 4. Nyquist and Bode plots of EIS spectra for blank CuNi alloy in different seawater flow rates. Dots present experimental data and lines fitted data obtained by fitting the spectra to models given in Fig. 5.

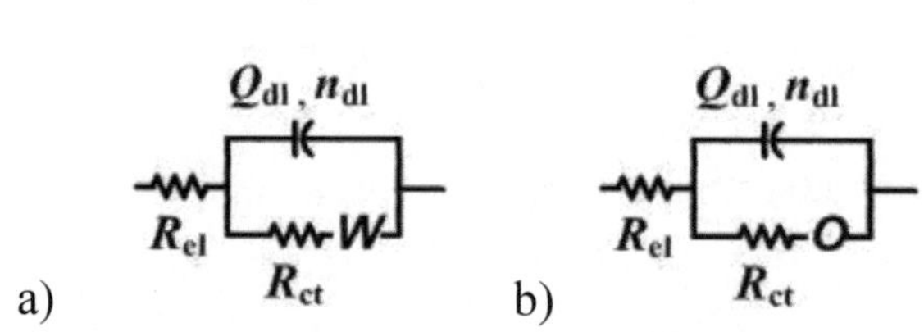

Figure 5. Equivalent electrical circuits used for fitting the EIS spectra of untreated sample.

Table 1. Impedance parameters obtained by fitting experimental data to selected equivalent circuits for untreated sample.

SAMPLE	Q_{dl} / μS sec^n	n_{dl}	R_{ct} / kΩ cm^2	W / μS sec$^{0.5}$	Q / μS sec$^{0.5}$	O / sec$^{0.5}$
0 L/min	177.8	0.72	0.90	9017	-	-
0.8 L/min	215.7	0.73	0.69	11620	-	-
1.6 L/min	257.1	0.72	0.58	12870	-	-
2.4 L/min	266.6	0.73	0.56	-	11190	2.42

It can be observed that all impedance parameters for blank CuNi sample change with the change of the flow rate (except n_{dl}). While double layer capacitance increases with an increase of the flow rate, charge transfer resistance and diffusion impedance decrease. This is a clear confirmation of the improved mass transfer of corrosive species, like O_2 and Cl^- towards the metal surface.

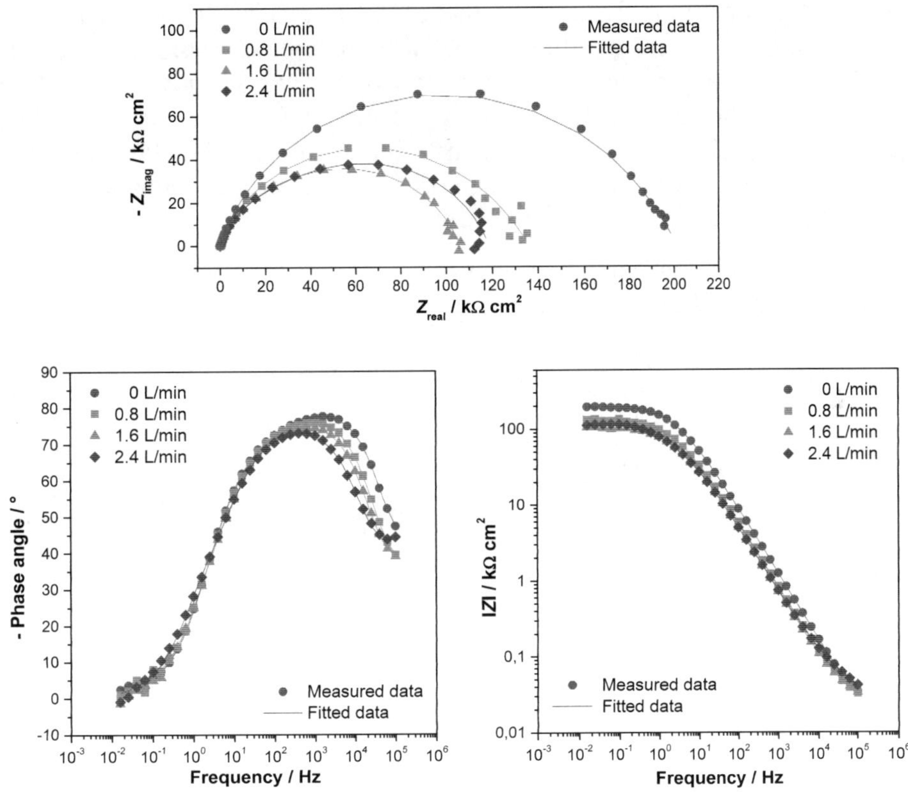

Figure 6. Nyquist and Bode plots of EIS spectra for treated CuNi alloy in different seawater flow rate. Dots present experimental data and lines fitted data obtained by fitting the spectra to models given in Fig. 7.

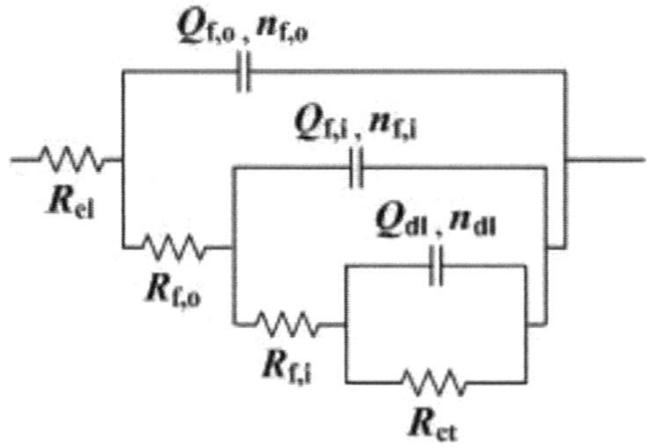

Figure 7. Equivalent electrical circuits used for fitting the EIS spectra of treated sample.

Impedance spectra obtained for ODPA treated samples are given in Figure 6. Generally, impedance decreases with the increase of the flow rate, although at the 2.4 L/min impedance is slightly bigger than for the 1.8 L/min. Obtained spectra was possible to model with an equivalent electrical circuit presented in Fig. 7. It consists of 3*RQ* combinations, where the one at the highest frequencies represents the capacitance and the resistance of the outer ODPA film ($Q_{f,o}$ and $R_{f,o}$), at medium frequencies the capacitance and the resistance of the inner ODPA film ($Q_{f,i}$ and $R_{f,i}$). The obtained impedance parameters are given in Table 2. In general, capacitive values increase with increase of the flow rate indication decrease of the inhibitor film thickness. Interestingly, the film resistance was the highest for the highest flow rate. On the other hand, charge transfer resistance decreased at higher flow rates similarly to what was seen for blank samples. However, the obtained resistance values for treated samples are much higher than those observed for the blank sample

Table 2. Impedance parameters obtained by fitting experimental data to selected equivalent circuits for ODPA treated sample.

SAMPLE	$Q_{f,o}$ / µS sec^n	$n_{f,o}$	$R_{f,o}$ / kΩ cm^2	$Q_{f,i}$ / µS sec^n	$n_{f,i}$	$R_{f,i}$ / kΩ cm^2	Q_{dl} / µS sec^n	n_{dl}	R_{ct} / kΩ cm^2
0 L/min	0.34	0.84	0.07	0.06	0.96	44.35	0.57	0.60	156.70
0.8 L/min	0.38	0.82	0.04	0.25	0.91	38.33	1.28	0.56	97.75
1.6 L/min	0.69	0.79	0.06	0.25	0.91	63.25	2.71	0.74	42.68
2.4 L/min	0.60	0.78	0.09	0.38	0.88	68.01	3.39	0.79	50.17

Influence of flow of medium in time

In order to determine the longevity of the protection at a constant flow rate, the samples were tested in time by polarization measurements in the narrow range of potentials, and the polarization resistance dependence on time of the exposure to river water is shown in Figure 8.

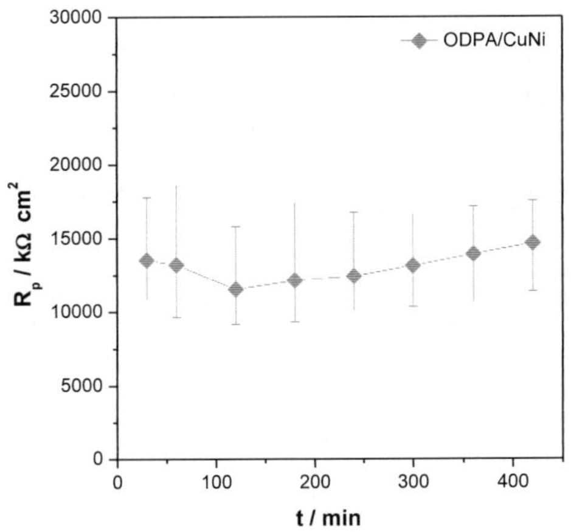

Figure 8. Polarization resistance dependence on time of exposure to river water measured at a flow of 2.4 L/min.

From Fig. 8., it can be seen that although there is a deviation from the mean value deviation occurs during the 7 h test at the highest tested media flow rate, the ODPA film is stable and does not result in loss of its protective properties.

Conclusion

The results obtained in this work show that the protective films of ODPA can efficiently protect copper-nickel and stainless steel from the corrosion in flowing natural waters (river and sea water).

Acknowledgments: The research leading to these results has received funding from Croatian Science Foundation under grant agreement 9.01/253.
The financial support of the Foundation of the Croatian Academy of Sciences and Arts, for projects "Corrosion protection of metals in natural waters" is gratefully acknowledged.

References

1. B.M. Silverman, K.A. Wieghaus, J. Schwartz, Langmuir, **21** (2005) 225.
2. R. Quinones, A. Raman, E.S. Gawalt, Thin Solid Films, **516** (2008) 8774.
3. M. Dubey, T. Weidner, L.J. Gamble, D.G. Castner, Langmuir, **26** (2010) 14747.
4. T. Hauffman, A. Hubin, H. Terryn, Surf. Interface Anal., **45** (2013) 1435.
5. A. Raman, M. Dubey, I. Gouzman, E.S. Gawalt, Langmuir, **22** (2006) 6469.
6. E. Stupnišek-Lisac, N. Galić, R. Gašparac, Corrosion, 56 (2000) 1105.

Influence of the Ni/Co ratio and of the Cr content on the high temperature oxidation of cast {Ti or Ta}-rich (Ni & Co)-based alloys

Patrice BERTHOD[1], Lydia TOUBAL[2], Zohra HIMEUR[2], Ouarda ABED[2], Mélissa LÉGLISE[1]

[1] *University of Lorraine / Institut Jean Lamour, France*

[2] *Faculty of Sciences and Technologies, France*

Melting and solidifying alloys in metallic mold under inert atmosphere allow introducing reactive elements and obtaining fine microstructures. A first series of alloys, based on Ni and Co with different ratios, containing 25wt.%Cr, 0.4wt.%C and either 6wt.%Ta or 1.6wt.%Ti, was synthesized by this way to study their microstructures and their oxidation behavior at high temperature (1127 and 1237°C). A second series of alloys for specifying the specific effect on the carbides population and on the oxidation at H.T. of a Cr lowering in {6wt.%Ta, 0.4wt.% C}-containing Ni-based alloys was also considered.

Charges made of pure elements were melted by high frequency induction heating under 300 mbars of pure argon. Rapid solidification was carried out in a cooled copper crucible. The microstructures of the as-cast alloys were characterized by XRD, SEM and EDS. Other samples were prepared and tested in oxidation by air for 20-24 hours at 1127°C and 1237°C. Cross-sections allowed characterizing their HT oxidation behaviors by specifying the oxidation products formed, as well as the chemical and microstructural modifications induced in subsurface.

In all cases rapid solidification resulted in fine microstructures. Primary carbides formed, chromium carbides and tantalum carbides, or chromium carbides and titanium carbides, with TaC/Cr_xC_y and TiC/Cr_xC_y surface fraction ratios increasing with the increase in Co content at the expense of the Ni one. At the same time, the behavior in oxidation by air at 1127°C or 1237°C, initially very good for the Ni-based Co-free alloys (chromia-forming) became more and more deteriorated. The worst behavior was observed for the Co-based Ni-free alloys.

The study of the second series of alloys showed that decreasing the Cr content induced a progressive disappearance of all carbides instead the stabilization of TaC as initially expected. Logically the oxidation resistance was progressively lost.

Behavior of three cast chromium-rich cobalt-based alloys in oxidation at 900°C in a 100% water vapour atmosphere

Patrice BERTHOD[1], Mélissa RITOUET[2], Thierry SCHWEITZER[3], Lionel ARANDA[4], Kevin GINESTAR[5], Sophie BOSONNET[6]

[1]*Institut Jean Lamour (UMR CNRS 7198), dept CP2S, University of Lorraine, B.P. 70239, 54506 Vandoeuvre-lès-Nancy, France, E-mail: patrice.berthod@univ-lorraine.fr*
[2]*Institut Jean Lamour (UMR CNRS 7198), dept CP2S, University of Lorraine, B.P. 70239, 54506 Vandoeuvre-lès-Nancy, France, E-mail: melissa.ritouet@univ-lorraine.fr*
[3]*Institut Jean Lamour (UMR CNRS 7198), dept CP2S, University of Lorraine, B.P. 70239, 54506 Vandoeuvre-lès-Nancy, France, E-mail: thierry.schweitzer@univ-lorraine.fr*
[4]*Institut Jean Lamour (UMR CNRS 7198), dept CP2S, University of Lorraine, B.P. 70239, 54506 Vandoeuvre-lès-Nancy, France, E-mail: lionel.aranda@univ-lorraine.fr*
[5]*CEA Saclay, DEN- DANS- DPC- SCCME- LECNA, Bât 458 P118, PC 50, 91191 Gif sur Yvette Cedex, France, E-mail: kevin.ginestar@cea.fr*
[6]*CEA Saclay, DEN- DANS- DPC- SCCME- LECNA, Bât 458 P118, PC 50, 91191 Gif sur Yvette Cedex, France, E-mail: sophie.bosonnet@cea.fr*

Abstract: In contrast with nickel-based and iron-based superalloys, the cobalt-based ones were not often studied in high temperature oxidation in presence of water vapour, notably in high concentration of H_2O. In this work three cobalt-based alloys or superalloys elaborated by foundry, of an increasing complexity, were subjected for 46 hours and 100 hours to oxidation at 900°C in a resistive furnace in a flow of pure water vapour (but presence of almost 5×10^{-5} atm. of O_2). The first alloy was a simple single-phased Co-10Ni-30Cr (wt.%) ternary alloy. The second one contained additionally 0.5C and was consequently double-phased with the presence of an interdendritic network of eutectic chromium carbides. The third alloy contained 7.5Ta, this leading to an interdendritic network of tantalum carbides instead the chromium carbides of the second alloy. Whatever the alloy and the oxidation duration, significant oxide spallation occurred in all cases. XRD was performed on the oxidized surfaces and showed that, over the ternary CoNiCr alloy, chromia was mainly present after 46h (presence of spinel $(Co,Ni)Cr_2O_4$ too) and the single oxide present after 100h. The quaternary CoNiCrC alloy was covered only by chromia, whatever the duration, and the $CrTaO_4$ was systematically detected together with chromia on the oxidized quinary CoNiCrCTa alloy (with also NiO only after 46h). All the three alloys were depleted in chromium in the subsurface (26, 24 and 17Cr in extreme surface for the CoNiCr, CoNiCrC and CoNiCrCTa alloys respectively). The carbides close to the alloy/oxide interfaces as well as the microstructures of the two carbides-containing alloys were not affected by the exposures at 900°C (temperature and time both not high enough), in contrast with what usually occur for these alloys at 1200 and even 1100°C. To conclude, the three alloys remained all chromia-forming during this test, and remain stable, on surface as well as in the bulk. New tests are envisaged to see if this remains true for longer durations. The main problem is the severe oxide spallation which took place.

Keywords: Cobalt-based alloys; Chromium carbides; Tantalum carbides; High temperature oxidation; Water vapour;

Introduction

Beside the famous single-crystalline γ/γ' superalloys other metallic alloys retain high interest for some high temperature applications. This is for example the case of the foundry-made cobalt-based superalloys family [1]. These alloys may offer to the users high mechanical strength in service at elevated temperature as well as correct resistance against oxidation by gases and corrosion by molten salts [2]. A typical example of use is as the constitutive material of glass working tool for shaping molten glass [3]. The oxidative gaseous milieu in which cast cobalt-based superalloys work the most often are dry or wet air, or mixtures of burner gases. However there are applications, notably in the nuclear power field, in which hot gases may contain very high quantities of water vapour and for which cobalt-based superalloys may be considered. Their behaviour in such water-rich gaseous environment seems to be never investigated and this is the reason why testing a little series of cast cobalt-based alloys in pure water vapour was decided here.

This work was carried out in the frame of the research activities of the GDR N°3583 "EVACOHT" (*Groupe De Recherche "Effet de la Vapeur d'eau en Atmosphères complexes sur la Corrosion et l'Oxydation à Haute Température*), of the CNRS (*Centre National de la Recherche Scientifique*).

Experimental

Materials

The cobalt alloys of interest in this study are three. The first one is a Co(bal.)-10Ni-30Cr (all contents in wt.%) ternary alloy, expected to be single-phased. The second one is the same as the preceding one but with addition of carbon (0.5wt.%C) to form chromium carbides. The last one is the previous alloy to which tantalum was added (7.5wt.% Ta) to promote the formation of tantalum carbides (TaC) exclusively. These three alloys were elaborated from pure elements, by high frequency induction foundry under inert atmosphere, by the Institut Jean Lamour laboratory.

Tests and first characterization

The three alloys were exposed in a 100% H_2O atmosphere at 900°C by the CEA/LECNA laboratory: a sample per alloy during 46 hours and a sample per alloy during 100 hours. After test, the samples were weighed for observing possible lass variation during the whole high temperature exposure and first macroscopic examinations were carried out, by CEA/LECNA.

Deeper characterization

Back to the Institut Jean Lamour laboratory, the tested samples were subjected to the characterization of their surfaces by X-ray diffraction (XRD) and Scanning Electron microscopy (SEM). After cross-cutting, embedding and grinding/polishing the obtained cross-sectional samples were subjected to the analysis of the corrosion products present on surface, and to the characterization of the sub-surface changed by the oxidation phenomena and of the bulk possibly modified by the exposure to high temperature. This was achieved by using SEM examinations and Energy Dispersive Spectrometry (EDS) measurements.

The flow diagram of the samples is illustrated in Figure 1 while the VAPOX apparatus is schematized in Figure 2.

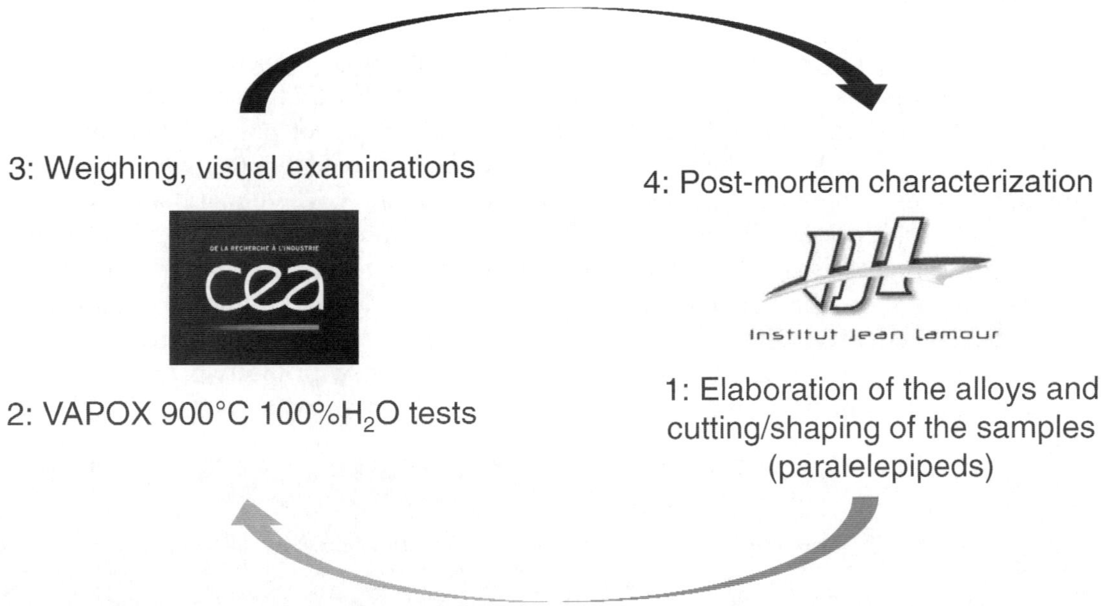

Figure 1: Sample path between Institut Jean Lamour and CEA/LECNA

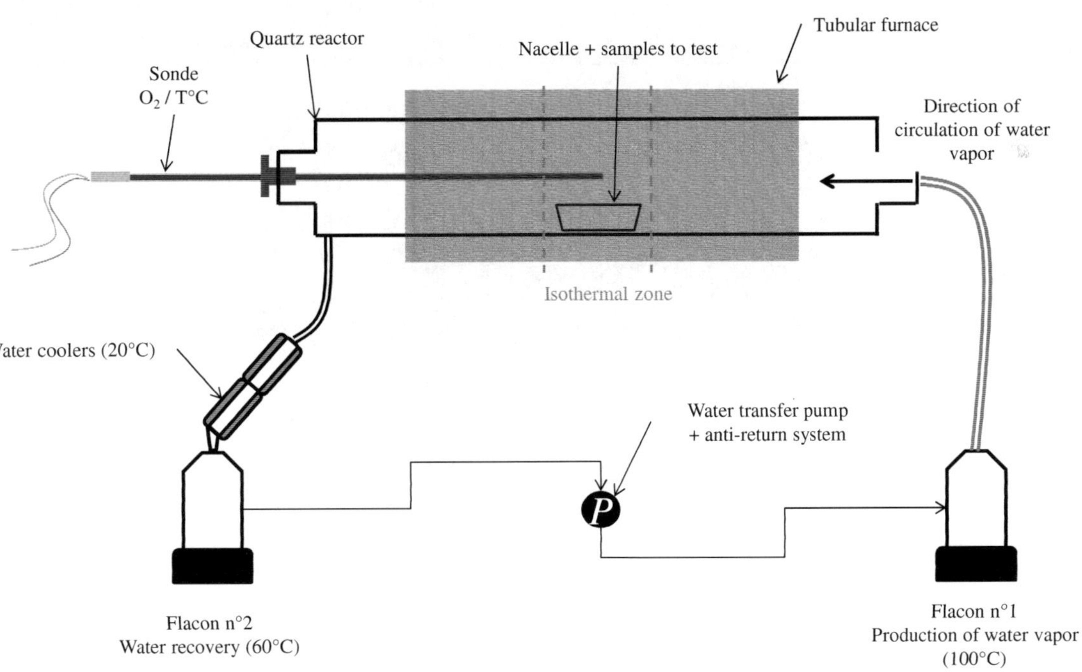

Figure 2: Description of the VAPOX apparatus (CEA/LECNA)

Results and discussion

Alloys: chemical compositions and microstructures before the high temperature tests

The chemical composition of the three alloys, controlled by EDS measurements, were obviously well respected for Co, Cr and Ta. In contrast it was not possible to specify the carbon content, due to its low molar weight and its low weight content in the two last alloys. The microstructures of the three alloys are illustrated in Figure 3. The ternary alloy seems to be single-phase, as expected. The second alloy is double-phased, with the co-existence of a dendritic matrix of Co-based solid solution and of interdendritic eutectic chromium carbides (black particles). The third alloy contains the Co-based solid solution matrix and interdendritic eutectic tantalum carbides (white particles), as major phases. Some black particles are also present, in much lower quantities compared with the two former phases; they are probably chromium carbides.

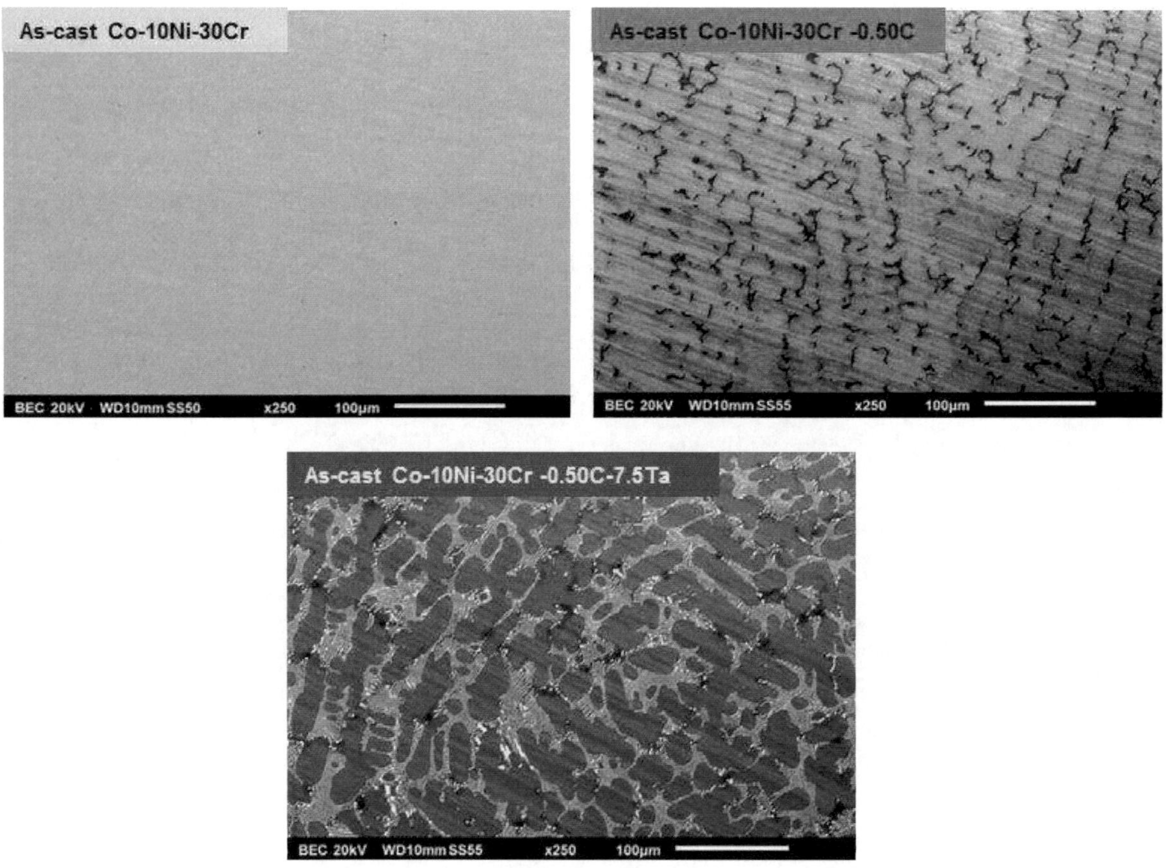

Figure 3: Microstructures of the alloys in their as-cast states

Visual examination of the samples after the high temperature tests

The macroscopic aspects before and after tests of the samples can be seen in Figure 4 for the 46h-exposures and in Figure 5 for the 100h-exposures. Macrographs (higher magnification than above) of the three exposed alloys for the two durations are given in Figure 6 to observe the oxidized state in better conditions. It appears that the external scales have more or less spalled off during the post-isothermal exposure cooling, and by considering the colour of some places along the nacelle, that chromia probably formed over all samples. Spallation was also confirmed by the negative values of mass variation observed for 5 out of the 6 samples.

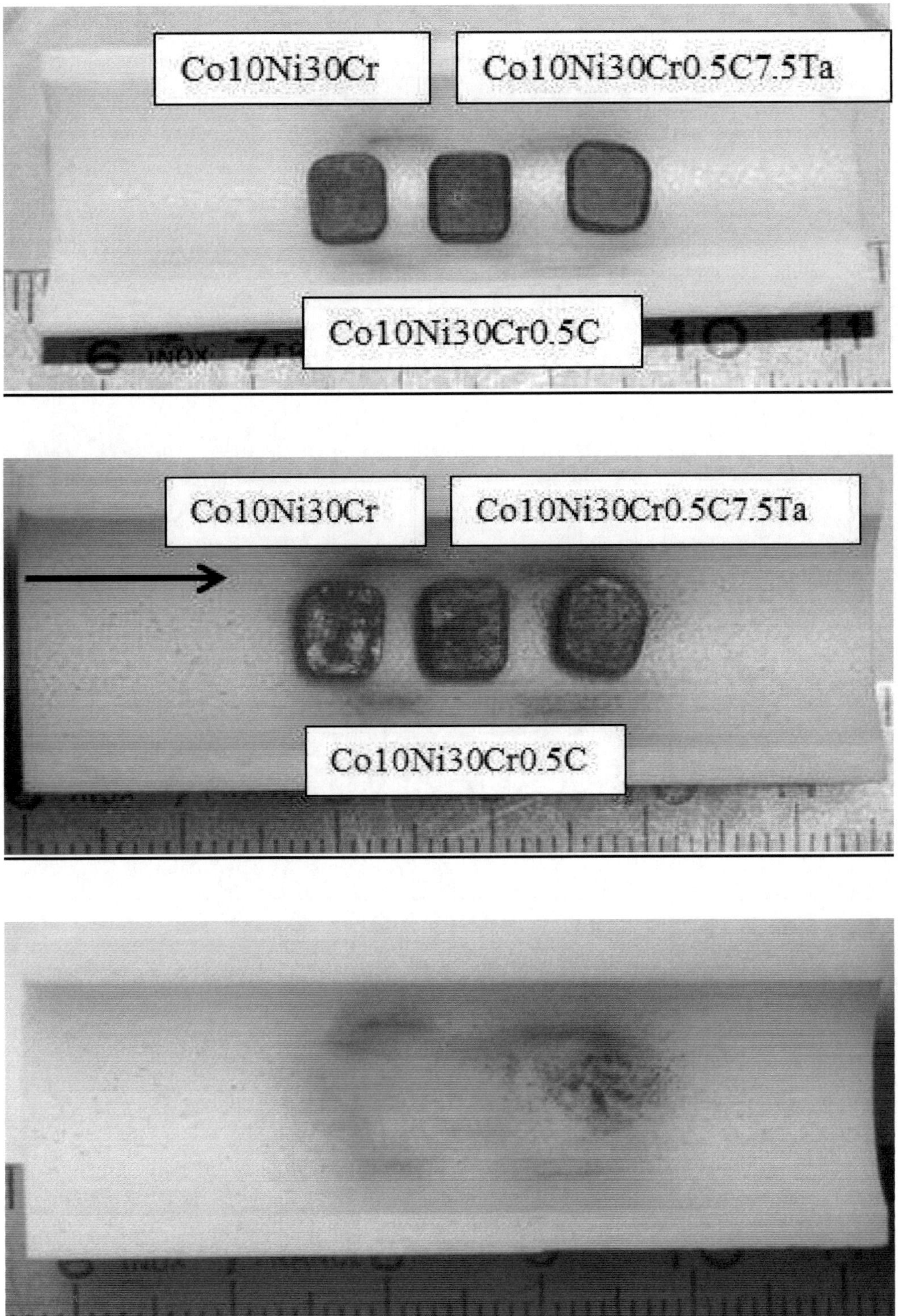

Figure 4: Aspects of the samples before (top) and after (middle position) 46h spent in 100%H_2O at 900°C; nacelle state after exposure (bottom)

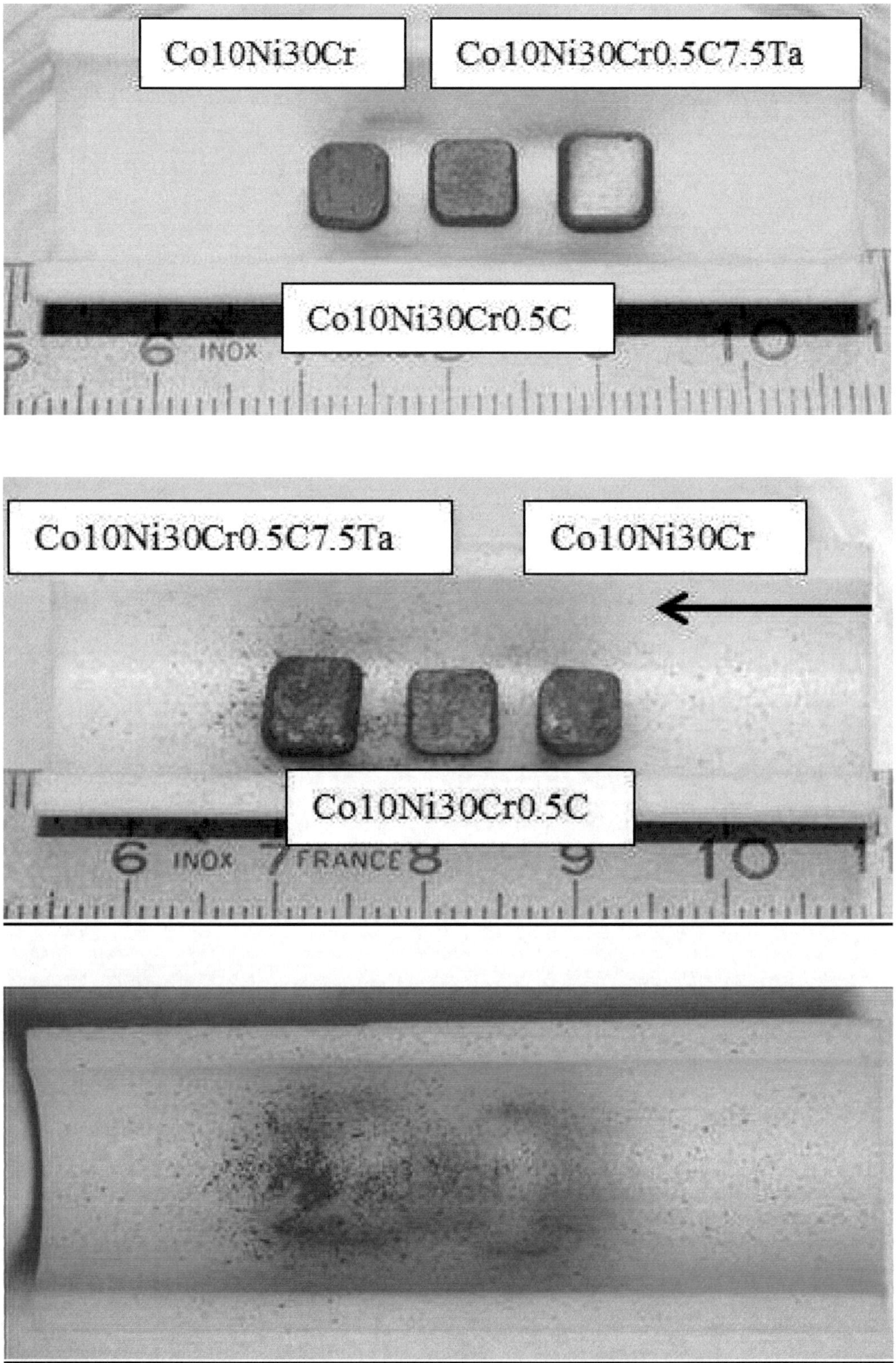

Figure 5: Aspects of the samples before (top) and after (middle position) 100h spent in 100%H_2O at 900°C; nacelle state after exposure (bottom)

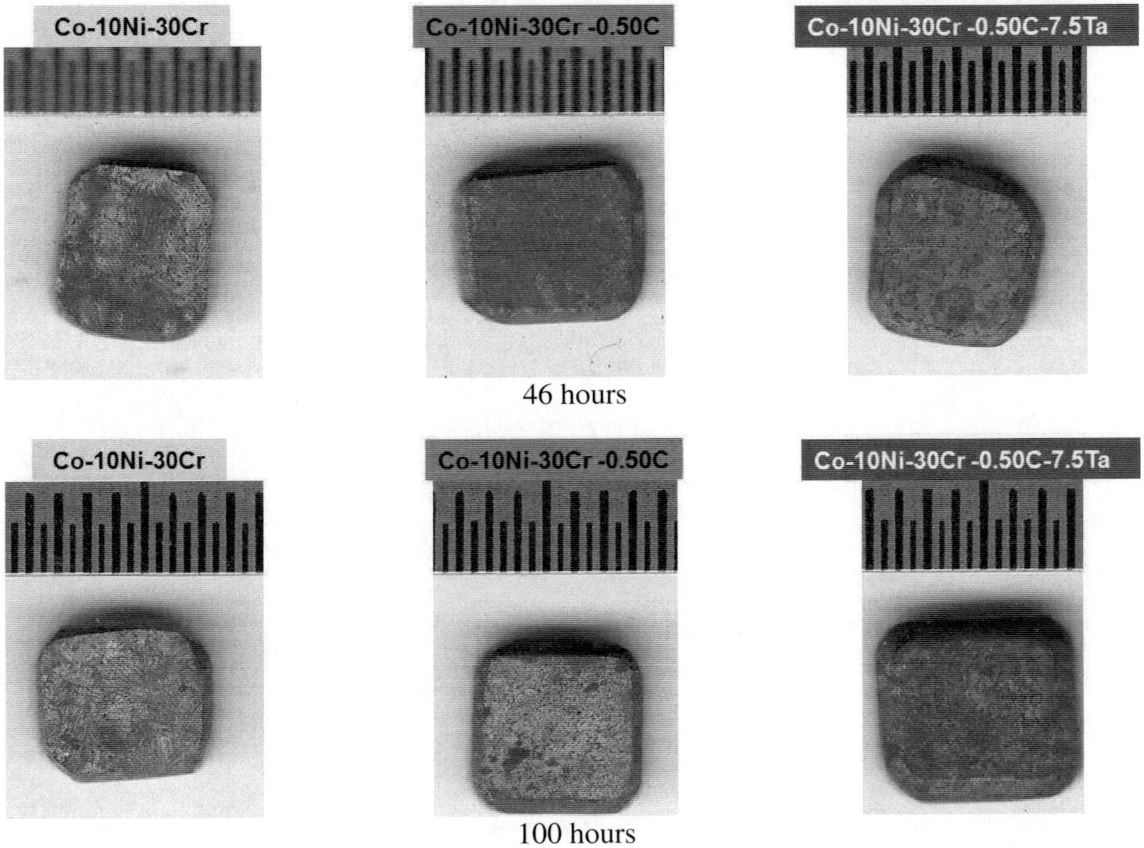

Figure 6: Macrographs of the samples after 46h (top) or 100h (bottom) in 100%H_2O at 900°C

Results of X-ray diffraction

The presence of chromia in all cases (3 alloys × 2 durations) is confirmed by the diffractograms obtained on the oxidized surfaces before cross-sectional preparation (Figure 7 for the Co-10Ni-30Cr alloy, Figure 8 for the 0.5C-containing alloy and Figure 9 for the 7.5Ta-containing one). The third alloys really behaved as chromia-forming alloys. However, one can notice also the presence of the MCr_2O_4 spinelle oxide (M=Co and Ni) in the case of the ternary alloy oxidized during 46 hours only, and of the $CrTaO_4$ oxide (classically observed for Ta-containing Co-based or Ni-based alloys [4, 5]).

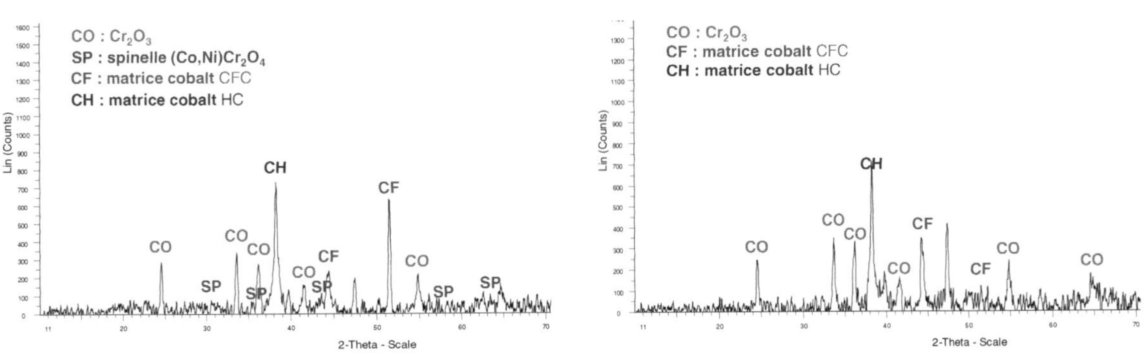

Figure 7: Diffractograms obtained for the Co-10Ni-30Cr alloy exposed to 100%H_2O at 900°C for 46 hours (left) and for 100 hours (right)

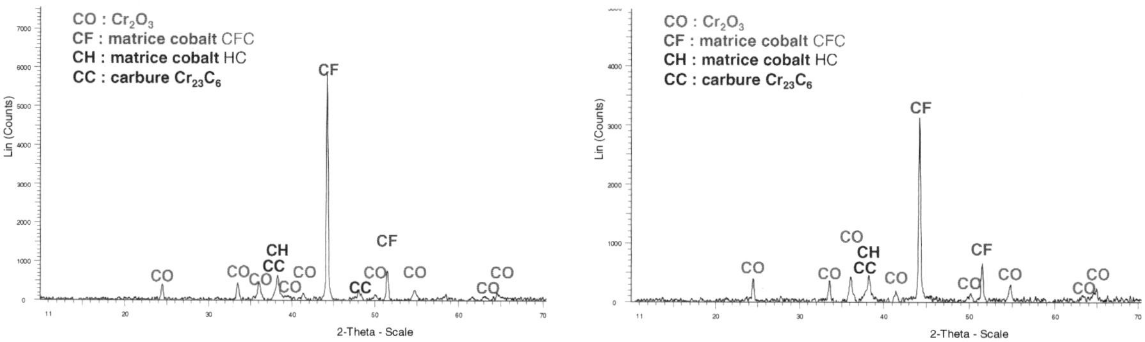

Figure 8: Diffractograms obtained for the Co-10Ni-30Cr-**0.5C** alloy exposed to 100%H_2O at 900°C for 46 hours (left) and for 100 hours (right)

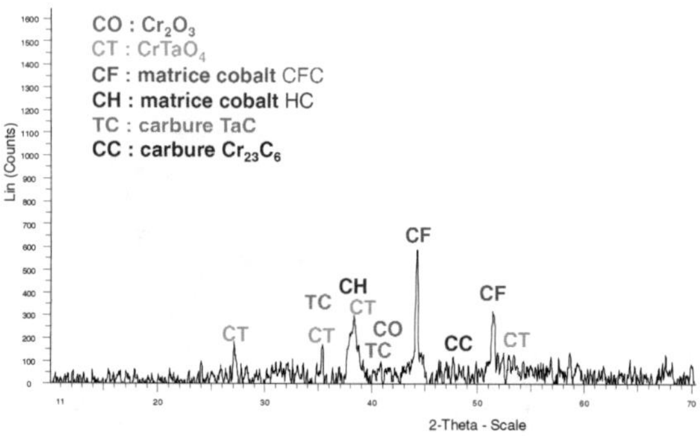

Figure 9: Diffractograms obtained for the Co-10Ni-30Cr-**0.5C-7.5Ta** alloy exposed to 100%H_2O at 900°C for 46 hours (left); right: diffractogram not yet available

The matrix, which is finally not single-phased but double-phased (Co-based solid solution of the HCP structure and of the FCC one simultaneously), is clearly evidenced by XRD, certainly more because the oxide spallation during cooling than because too thin external oxide scales. For the second alloy the chromium carbides are detected by XRD. This is also the case for the tantalum carbides for the third alloy.

SEM observation of the surface oxidized states

The surface states were observed at higher magnification using the SEM. Observations were carried out with two modes: Secondary Electrons (to enhance the relief irregularities on surface) and Back Scattered Electrons (to enhance the contrast of average atomic number). Some of the obtained micrographs are given in Figure 10 for the three alloys after 46 hours of exposure, and in Figure 11 for the 100 hours of exposure. One can confirm that more or less severe spallation occurred during the cooling since significant fractions of denuded alloy are visible (white in BSE mode). One also feels that spallation was maybe more severe after 100 hours of exposure than after only 46 hours. Even if they are also visible in the areas having still kept their external oxide scale, the stripes issued from the last grinding of the samples to be tested (SiC paper @ #1200 grit), the grinding stripes are clearly evidenced on the denuded parts of the samples. Oxidation was thus never strong enough to erase them. For the carbides-containing alloys, on can additionally notice that the chromium carbides are still emerging,

this demonstrating a first time that the 46h - and even also 100h – oxidation at 900°C are not long enough to allow a carbide-free depth developing from the oxide/alloy interface. In contrast, the tantalum carbides emerging on surface were oxidized, in $CrTaO_4$ (pale areas in BSE mode), in the case of the Co-10Ni-30Cr-0.5C-7.5Ta alloy. Since they are known to favour the spallation of the external chromia scale during cooling, and even during the oxide growth in isothermal conditions, it is logical to observe that spallation was the most severe for this alloy.
The cross-sectional examinations can be illustrated by the SEM/BSE micrographs presented in Figure 12. No external scale can be seen, these zones having been concerned by spallation at cooling. Close to the oxide (lost) / alloy interface, the subsurface is not really changed. Notably no real carbide-free zone developed enough to be seen (900°C is a too low temperature for such alloys). However it can be remarked that the part of the white TaC carbides emerging on surface are obviously converted in pale oxides.

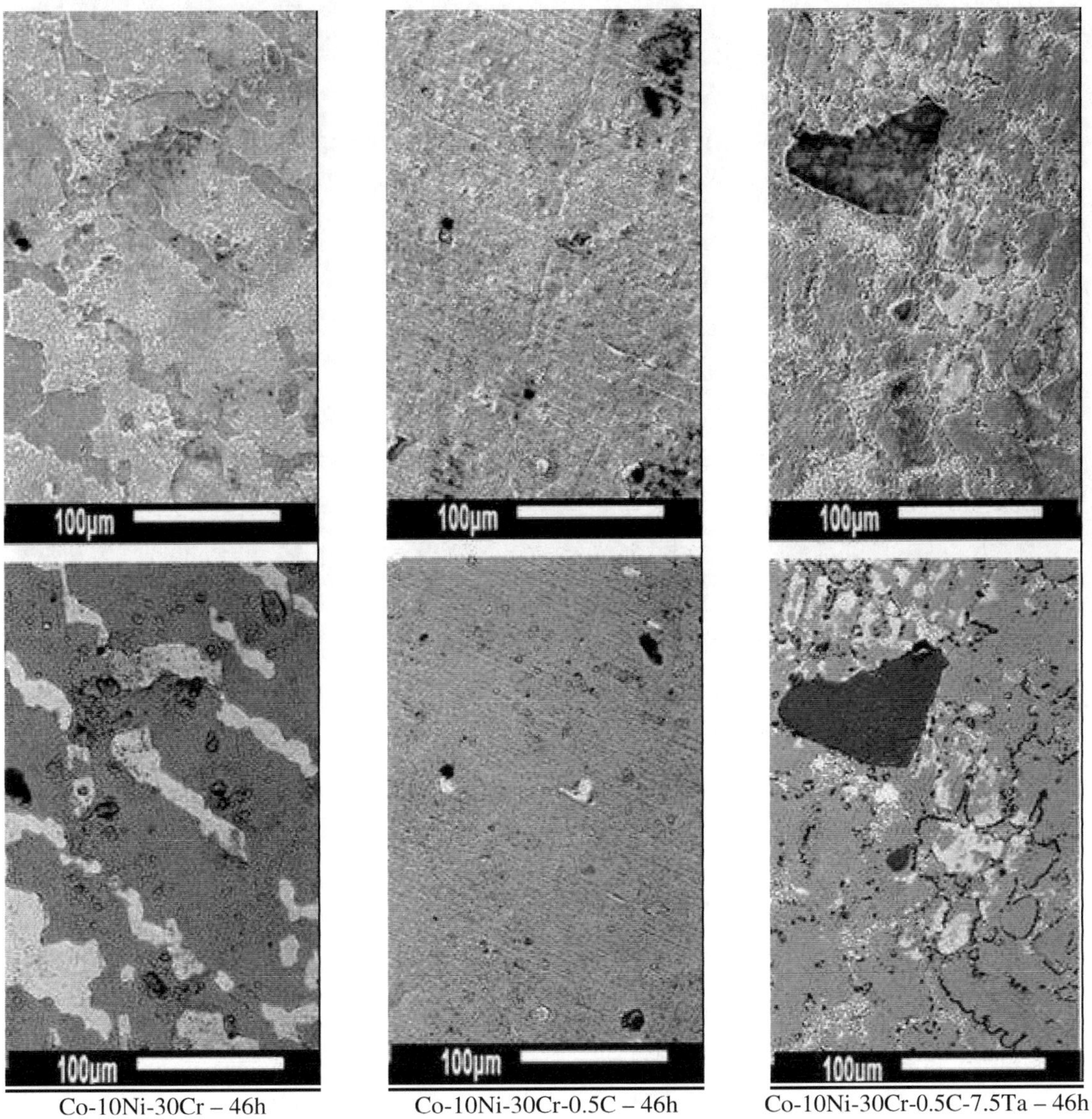

Figure 10: SEM micrographs taken on the oxidized surfaces of the three alloys exposed to 100%H_2O for 46 hours (top: Secondary Electrons mode, bottom: Back Scattered Electrons)

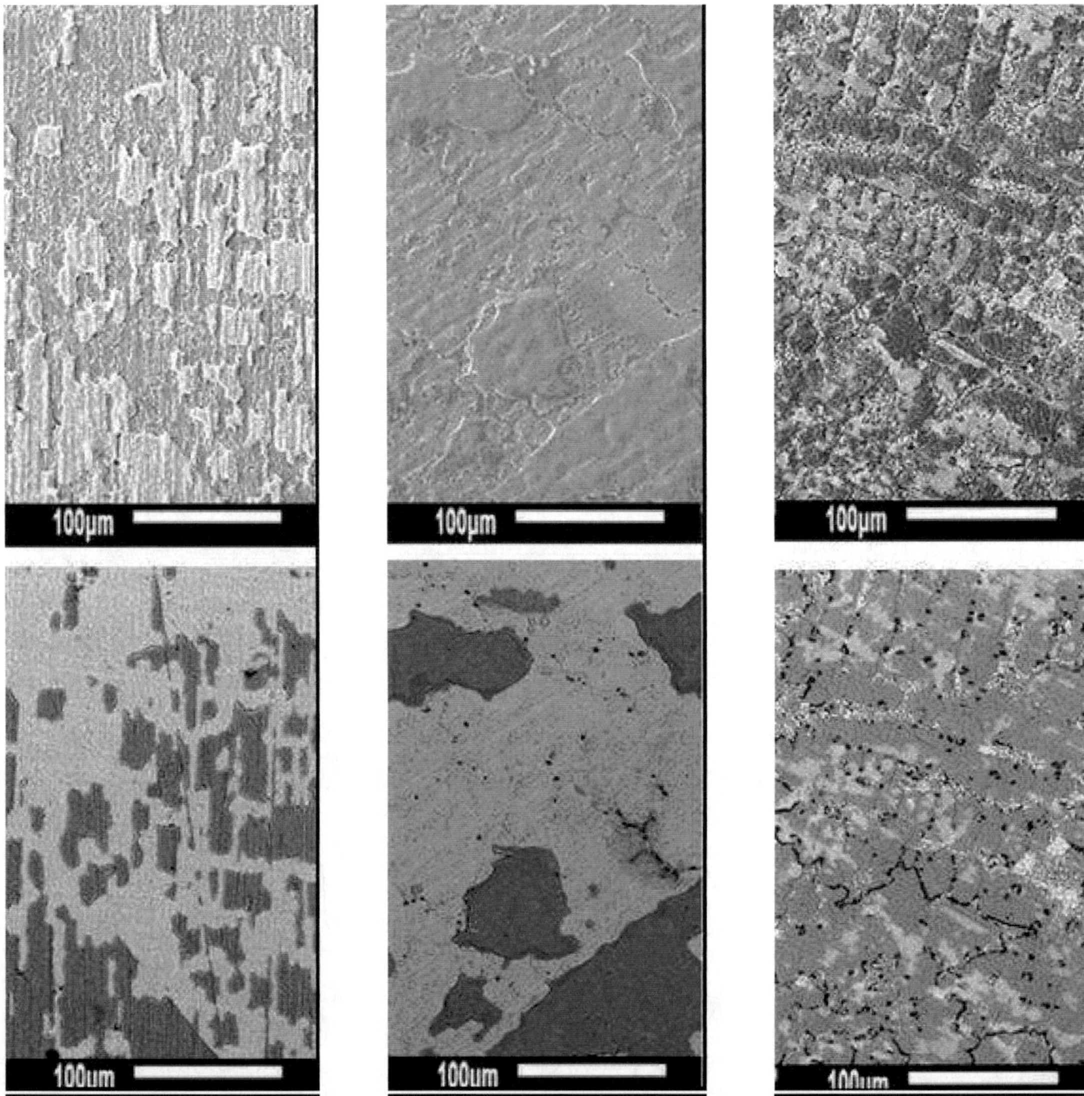

Figure 11: SEM micrographs taken on the oxidized surfaces of the three alloys exposed to 100%H_2O for 100 hours (top: Secondary Electrons mode, bottom: Back Scattered Electrons)

Some spot EDS analyses were carried out in the alloy close to the outer surface. The local chromium content was unsurprisingly decreased after high temperature oxidation but only down to about 26 wt.%Cr for the first alloy and 23 wt.% for the second one. The chromium content in extreme surface fell down to 17 wt.% for the third alloy. Thus, the two first alloys went on being chromia-forming but this behaviour was probably threatened in the case of the third alloy.

Microstructure changes in the bulk

The same micrographs can be also exploited to analyse the changes in the microstructure of the alloys induced by the exposure at 900°C. So, deeper in the alloys no change concerning the morphology of the carbides occurred. The chromium carbides remained acicular and the tantalum carbides kept their script-like shape; none of them was converted in alignments of globular carbides, as generally observed when the temperature of exposure is 1200 or 1100°C.

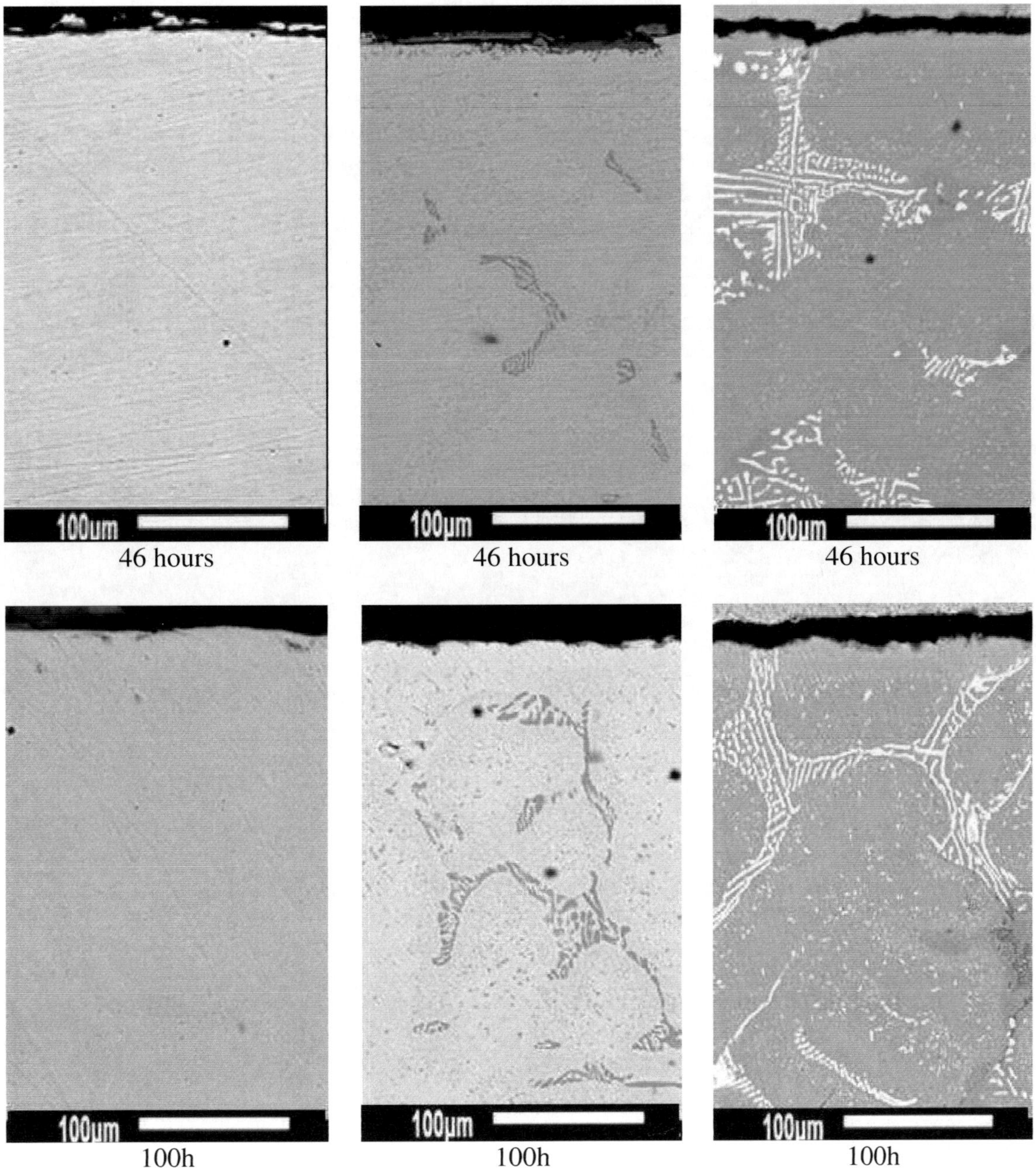

Co-10Ni-30Cr | Co-10Ni-30Cr-0.5C | Co-10Ni-30Cr-0.5C-7.5Ta

Figure 12: Cross-sectional SEM micrographs illustrating the surface and subsurface deterioration of the alloys after their thermal exposures

Far enough from the oxidation front, new carbides obviously precipitated in the matrix; these small and round secondary carbides are chromium carbides in the Co-10Ni-30Cr-0.5C alloy and TaC carbides in the third alloy. Such solid state precipitation is effectively possible in these types of alloys between 800 and 1000°C.

Conclusion

The three cobalt-based alloys appeared thus as rather resistant on long time against the atmosphere made of exclusively water vapour for this temperature of 900°C. Oxygen is particularly oxidant (it was present here too, in very small quantity) but water vapour is also known to be an oxidant gas. Water vapour is furthermore favourable to the volatilization of oxide species (unfortunately including the protective chromia and also alumina). This last aspect was not really studied here but may be later specified by using thermogravimetry techniques and by applying numerical treating methods [6, 7]. Sometimes cited for its beneficial effect on the resistance against spallation in thermal cycling, water vapour did not demonstrate this in the present study. It can be thought that such effect exists for some alloys different from the three present ones, or that too much water vapour may have the opposite effect. These are many questions which merit to be further investigated.

Acknowledgments

The authors thank the French national centre of the scientific research (CNRS, Centre National de la Recherche Scientifique*) for the funding which allowed the activities of the* Groupe De Recherche *N°3583 "EVACOHT".*

References

1. M. J. Donachie and S. J. Donachie, Superalloys: A Technical Guide, 2nd ed., ASM International, Materials Park, OH (2002).
2. D. J. Young, High Temperature Oxidation and Corrosion of Metals, Elsevier, Amsterdam, (2008).
3. P. Berthod, C. Liébaut and J.-L. Bernard, Cobalt based alloy, article made from said alloy and method for making same, international patent (WO/1999/016919).
4. P. Berthod and Z. Himeur, Materials and Corrosion, **69** (2018) 703.
5. P. Berthod, Metallurgical and Materials Transactions A, **49A** (2018) 2311.
6. P. Berthod, Oxidation of Metals, **64**(3/4) (2005) 235.
7. L. Aranda et al, Materials at High Temperature, **32**(5) (2015) 530.

Degradation at 550°C of cast chromium-rich cobalt-based alloys exposed to a H_2O-O_2-CO_2-HCl gaseous mixture

Lionel ARANDA[1], Thierry SCHWEITZER[2], Patrice BERTHOD[3], Christophe RAPIN[4], Didier SOUCHON[5], Farès MAAD[6], Didier SOUCHON[7]

[1]Institut Jean Lamour (UMR CNRS 7198), dep[t] CP2S, University of Lorraine, B.P. 70239, 54506 Vandoeuvre-lès-Nancy, France, E-mail: lionel.aranda@univ-lorraine.fr
[2]Institut Jean Lamour (UMR CNRS 7198), dep[t] CP2S, University of Lorraine, B.P. 70239, 54506 Vandoeuvre-lès-Nancy, France, E-mail: thierry.schweitzer@univ-lorraine.fr
[3]Institut Jean Lamour (UMR CNRS 7198), dep[t] CP2S, University of Lorraine, B.P. 70239, 54506 Vandoeuvre-lès-Nancy, France, E-mail: patrice.berthod@univ-lorraine.fr
[4]Institut Jean Lamour (UMR CNRS 7198), dep[t] CP2S, University of Lorraine, B.P. 70239, 54506 Vandoeuvre-lès-Nancy, France, E-mail: christophe.rapin@univ-lorraine.fr
[5]Veolia Environnement Recherche et Innovation, Zone Portuaire - 291 Avenue Dreyfous Ducas, 78520 Limay, France, E-mail: didier.souchon@veolia.com
[6]Veolia Environnement Recherche et Innovation, Zone Portuaire - 291 Avenue Dreyfous Ducas, 78520 Limay, France, E-mail: didier.souchon@veolia.com
[7]Veolia Environnement Recherche et Innovation, Zone Portuaire - 291 Avenue Dreyfous Ducas, 78520 Limay, France, E-mail: didier.souchon@veolia.com

Abstract: Cobalt-based superalloys may be considered as other refractory alloys based on nickel or iron for applications in waste-to-energy (WtE) boilers. Knowing their generally lower high temperature oxidation and corrosion resistance by comparison to nickel alloys for example, it can be informative to study how such cobalt alloys may behave in contact with hot atmospheres as complex as what can be met in an WtE boiler. In this work a rather complex cobalt-based alloy (Co-10Ni-30Cr-0.5C-7.5Ta) and a more simple Ta-free one (Co-10Ni-30Cr-0.5C) were synthesized by conventional casting. They were cut and prepared as polished samples, then exposed to a {18% H_2O, 8%O_2, 5%CO_2, 1100ppm HCl} gas mixture at 550°C during 260 hours, in a special lab-scale pilot, directly in the complex gaseous atmosphere for the first one, and immersed in synthetic ashes (NaCl) under the same complex gaseous atmosphere for the second one. After test the Ta-containing sample was subjected to X-ray diffraction and SEM observations. A {10 to15μm}-thick oxide scale is formed on surface. It seemingly grew both externally (probable cationic diffusion) and internally (probable anionic diffusion), as demonstrated by the shape of the initial TaC carbides still existing in the most internal half of the oxide thickness (still TaC in the alloy part but tantalum/tantalum&chromium oxides in the internal part of the oxide scale). This oxide scale was not chromia but Co(Ni)O in the most external part of the scale and $CoCr_2O_4$ (spinel) in its most internal part. XRD also showed the presence of NaCl and residual $CaSO_4$, probably coming from previous ashes tests performed in the lab-scale. These observations are different from previous results obtained for the same alloy but oxidized for about 50 hours at 1000°C in synthetic dry air (external oxide made of chromia (Cr_2O_3), spinel ($CoCr_2O_4$) and $CrTaO_4$ films, development of a TaC-free zone in subsurface). Concerning the second alloy (TaC-free) immersed in ashes no protective oxide layer seemed developing (dissolution in the molten salts?) and an internal oxidation/chloridation of the chromium carbides network over a 500μm-depth from surface happened, maybe through an active oxidation mechanism.

Keywords: Cobalt-based alloys; Hot oxidation; Hot corrosion; Water vapour; Complex atmosphere;

Introduction

For applications at moderate temperature, even in presence of intense mechanical stresses and very aggressive fluids, alloys which are needed can be ones which may be different from the superalloys used as turbine blades in aeronautics or power generation, and cheaper than them. This is for example the case of high performance stainless steels, nickel alloys, or why not cobalt-based alloys [1]. These alloys may bring interesting mechanical strength and correct resistance against oxidation by gases and corrosion by molten substances [2]. Cobalt alloys have rather found new applications such as hot tools for working molten glass [3]. The oxidative gaseous milieu in which cast cobalt-based superalloys work the most often are dry or wet air, or mixtures of burner gases. However there are applications, notably in the waste-to-energy (WtE) plants, in which hot gases are present as mixtures of many very aggressive chemical substances (in WtE boilers), for which cobalt-based superalloys may be considered. Their behaviour in so highly hostile environment seems to have been never investigated. This is why one decided here to test some cast cobalt-based alloys in hot environments typical of waste valorisation for energy production.

This work was carried out in the frame of the research activities of the GDR N°3583 "EVACOHT" (*Groupe De Recherche "Effet de la Vapeur d'eau en Atmosphères complexes sur la Corrosion et l'Oxydation à Haute Température*), of the CNRS (*Centre National de la Recherche Scientifique*).

Experimental

Obtaining the two cobalt alloys for this work

Two cobalt alloys, detsined to be elaborated by foundry, were designed for this study. Based on Co, it was decided to introduce 10wt.% Ni in their chemical compositions to stabilize the crystalline structure in Face Centred Cubic, favourable to high strength and good ductility at all temperature. To allow the alloys correctly resisting oxidation and corrosion at high temperature 30wt.%Cr were also added. To strengthen the interdendritic spaces 0.5 wt.%C were introduced to allow the formation of eutectic carbides. In presence of only chromium as carbide-former element chromium carbides may be expected in the present situation. But if a fourth element such as tantalum is also present in the composition, TaC carbides may appear and efficiently reinforce the alloys at all temperatures. To finish, the alloys were two: a Co-10Ni-30Cr-0.5C one (containing chromium carbides) and a Co-10Ni-30Cr-0.5C-7.5Ta one (containing tantalum carbides).
These two alloys were elaborated under neutral atmosphere by high frequency (HF) induction foundry (Figure 1 left) and their microstructures were obtained as expected (Figure 1 right).

The experimental apparatus: description

The apparatus used in this work is schematized in Figure 2. It is designed to reproduce the conditions of a working WtE boiler is composed of a gas burner which allow obtaining the required gases (H_2O, CO_2) at the wished temperature and to introduce scoria in the flow of the produced combustion gases. Gases and ashes are homogenized in the combustion chamber. Ashes are introduced by a screw feeder, pure SO_2 comes from a compressed gas bottle and

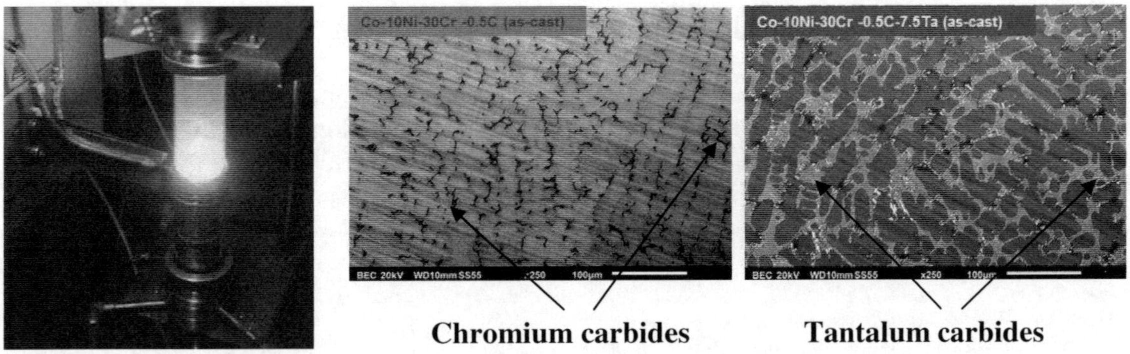

Figure 1: On-going elaboration of an alloy by HF induction melting (left) and SEM*/BSE** micrographs of the microstructures obtained for the two alloys (right)
**: Scanning Electrons Microscopy; **: Back Scattered Electrons mode*

HCl is introduced as an aqueous solution injected thanks to a metering pump. The produced smokes are mashed by a water shower and a decanter allows recovering the scoria.

The experimental apparatus: operating conditions

The temperature known by the samples is rated at 550°C. One of them, suspended, is exposed to the gas mixture (several hundred ppm HCl, several tens %.vol H_2O, about ten %.vol O_2 and CO_2) and the other one is immersed in a crucible containing synthetic ashes (NaCl). The test duration is 260 hours.

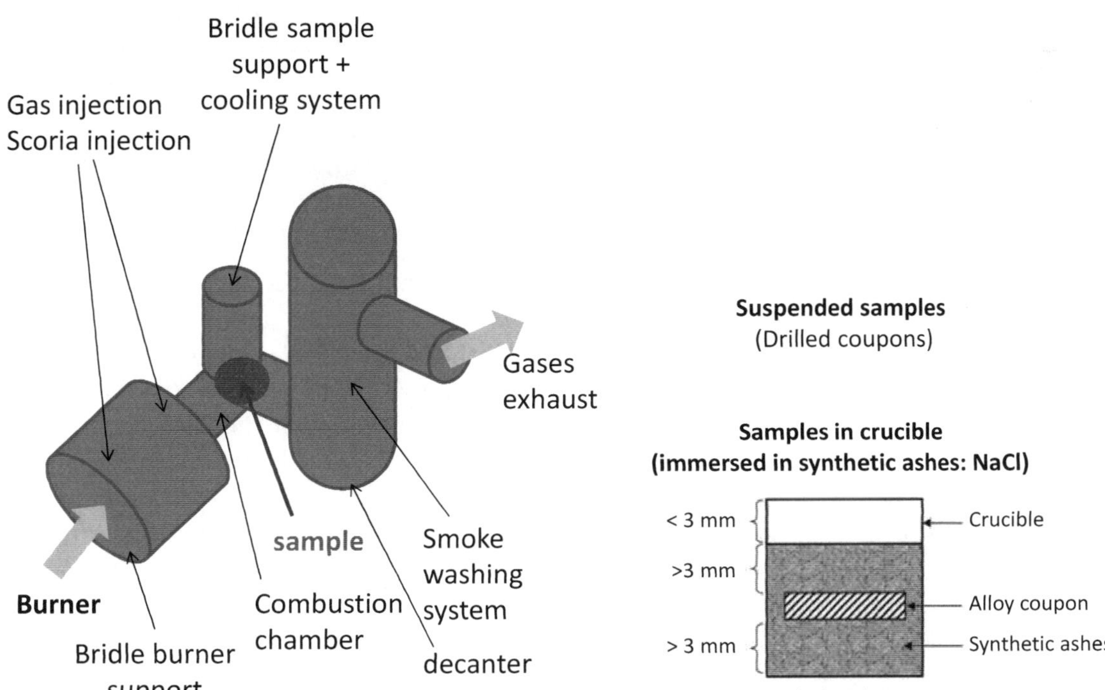

Figure 2: Scheme of the apparatus (left) and details of the environments of the two samples per alloy (the suspended one and the one immersed in the synthetic ashes, right)

Metallographic preparation of the tested samples and characterization

After test the samples were embedded in a cold resin mixture, cut and ground with SiC papers from 240 to 2400 grit, with Kerdane (dearomatized petrol, cleaning and degreasing) instead water as lubricant. Polishing was done using textile disk enriched with 1μm diamond particles, with lubrication with absolute ethanol. The two types of samples being prepared for observation are illustrated in Figure 3.

Sample which was suspended   Sample which was immersed

Figure 3: Macrophotos of the two types of samples being metallographically prepared

X-ray diffraction (XRD) runs were performed on the samples as well as electron microscopy observation (Scanning Electrons Microscopy, SEM, either in Secondary Electrons mode, SE, or in Back Scattered Electrons mode, BSE, and Energy Dispersive Spectrometry, EDS).

Results and discussion

Suspended samples: example of the Ta-containing alloy

SEM/SE micrographs of the surface of the suspended Co-10Ni-30Cr-0.5C-7.5Ta sample are given in Figure 4 (left), at low magnification and at higher magnification. The XRD results and EDS ones suggest the presence of NaCl and of $Ca(SO_4)$. This is rather curious for the suspended sample. What is thr origin of NaCl and of the calcium sulphate? The SEM/BSE micrographs (two magnifications again) presented in the right part of the same figure show the corrosion products as well as the subsurface degradation of the alloys. EDS spot analyses were carried out here and there in the layer of corrosion products. They evidenced the presence of the (Co, Ni)O oxide, the possible presence of the $CoCr_2O_4$ spinel oxide and of inclusion of $CaSO_4$ in the oxide scale.

The interdendritic network of eutectic TaC carbides seems unchanged, even close to the oxidation front. Obviously no carbide-free zone developed inwards and the script morphology of the TaC was not changed (no start of fragmentation notably). But it appears at high magnification that the oxidation developed inwards, consuming the alloy matrix and transforming the tantalum carbides in oxides with the same morphology.

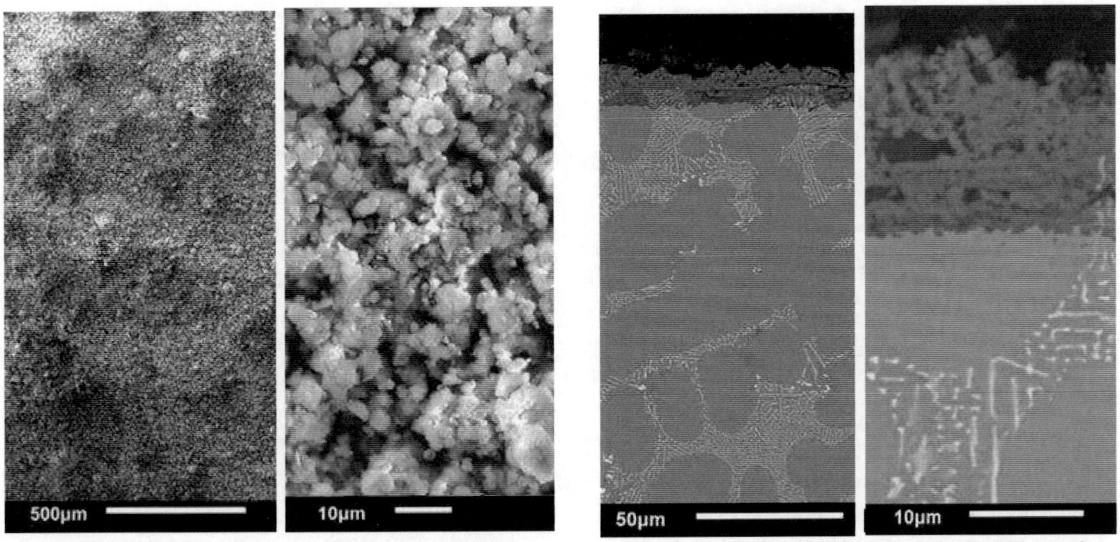

SEM images of the surface (SE mode) Cross sectional SEM images (BSE mode)

Figure 4: Surface SEM/SE observations (left) and cross-section SEM/BSE observation of the suspended Co-10Ni-30Cr-0.5C-7.5Ta alloy

Immersed samples: example of the Ta-free alloy

The Co-10Ni-30Cr-0.5C alloy was obviously severely corroded, as shown in Figure 5 in which the different layers formed over the alloy are identified. One notably points out the presence of an intermediate melting zone (Figure 6) in which the main constituent is NaCl with traces of Co and Cr. The presence of $CoCl_2$, and maybe also of $CrCl_3$ crystals was also detected by EDS. This suggests the formation of chlorides eutectic with low melting point.

Concerning the alloy itself, it significantly suffered from the immersion at high temperature in the synthetic ashes since no protective oxide scale (e.g. chromia) developed. Additionally internal corrosion occurred over a depth of more than 500 µm, with the oxidation / chloridation of the network of chromium carbides.

These observations evocate a process of corrosion involving similar phenomena as in the mechanism of active oxidation earlier described for steels [4].

Conclusion

Usually devoted to uses at temperatures much higher than the one applied in this work, the two considered cobalt-based alloys paradoxically suffered here much more than when they are exposed to air or in contact with molten glasses for example at temperatures higher than 1000°C. It is true that the 550°C that they knew here were much lower than what is required to allow a significant development of a carbide-free zone or fragmentation and coarsening of the interdendritic carbides, chromium carbides or TaC. But in the present context the burner

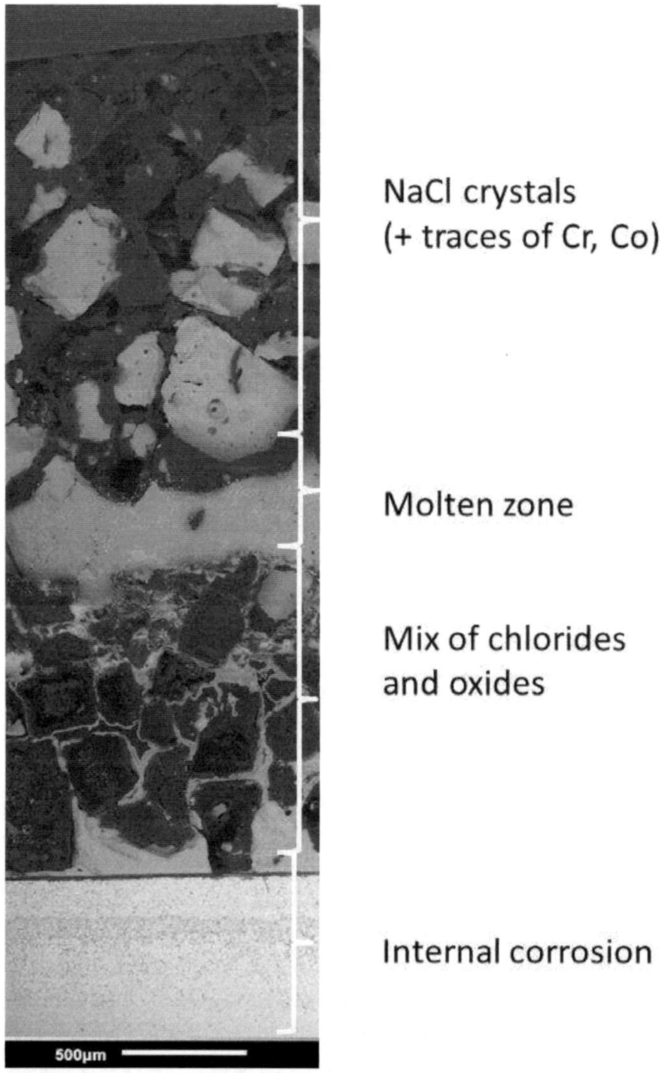

Figure 5: SEM/BSE observation of the corrosion products formed on the surface of the immersed Co-10Ni-30Cr-0.5C alloy (general view)

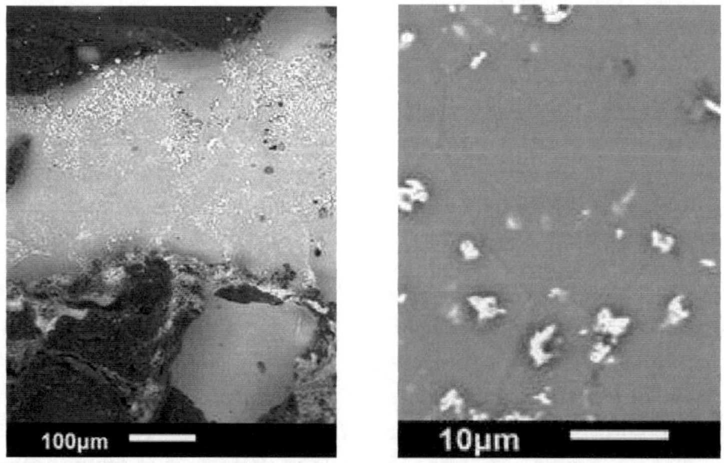

Figure 6: SEM/BSE observation of the corrosion products formed on the surface of the immersed Co-10Ni-30Cr-0.5C alloy (enlargements in the molten zone)

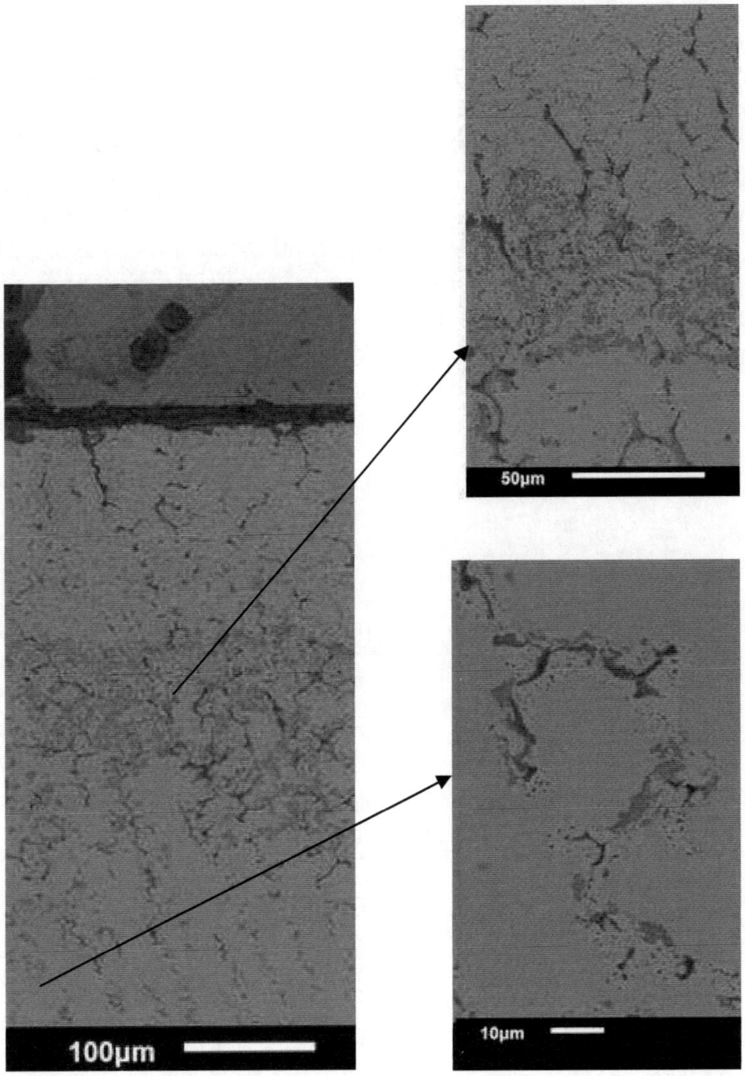

Figure 7: Internal corrosion of the Co-10Ni-30Cr-0.5C alloy (oxidation / chloridation of the network of chromium carbides over a depth of more than 500 µm)

gases and the synthetic ashes acted as redoubtable corrosive fluids against which chromia did not succeeded in forming to obstruct the induced deteriorating phenomena. After these factual descriptions of the degradations, next work will aim to understand what happened for the suspended sample as well as for the immersed ones. Some starts of explanations, extrapolated from the active oxidation in the case of steels, were suggested in this work but now they need to be deepened and completed.

Acknowledgments

The authors thank the French national centre of the scientific research (CNRS, Centre National de la Recherche Scientifique*) for the funding which allowed the activities of the* Groupe De Recherche *N°3583 "EVACOHT".*

References

1. M. J. Donachie and S. J. Donachie, Superalloys: A Technical Guide, 2nd ed., ASM International, Materials Park, OH (2002).
2. D. J. Young, High Temperature Oxidation and Corrosion of Metals, Elsevier, Amsterdam, (2008).
3. P. Berthod, C. Liébaut and J.-L. Bernard, Cobalt based alloy, article made from said alloy and method for making same, international patent (WO/1999/016919).
4. H. J. Grabke, Incinerating Municipal and Industrial Waste, Ed. Bryers R. W., Hemisphere Publ. Corp., 161-177 (1991).

Theoretical Study of Self-Assembled Monolayer of Imidazolines on Iron Oxide

Aurelie GOURON[1], Katell LE MAPIHAN[2], Sheyla CAMPEROS[3], Ahmad AL FARRA[3], Virginie LAIR[1], Armelle RINGUEDÉ[1], Michel CASSIR[1], Boubakar DIAWARA[1]

[1] *Chimie ParisTech, PSL Research University, CNRS, Institut de Recherche de Chimie Paris (IRCP), France*

[2] *TOTAL Research & Technology Feluy, Belgium*

[3] *Total Research and Technology Gonfreville, France*

Film-forming amines, as oleic imidazoline, prevent acid corrosion of carbon steel in head of atmospheric distillation process of petroleum refining. The commonly assumed inhibition mechanism is the formation of a self-assembled monolayer (SAM) of inhibitors molecules on the surface. The hydrophilic head group of imidazoline derivatives bind strongly on the surface and their alkyl chains forms a hydrophobic barrier above the surface which prevent corrosion.

Adsorption of SAM of different imidazolines derivatives was investigated on Fe_2O_3 (0001) surface with DFT calculations to understand their corrosion inhibition mechanism on carbon steel [1]. Quantum chemistry calculations enable us acquiring a better insight on the nature of imidazoline/surface interactions, i.e. partly covalent bonds, and self-organization of imidazoline ring on the surface than the previous force field-based studies [2].

DFT calculations have shown that the imidazoline ring and its pendent group adsorbs strongly on the surface and are linked to the surface through partly covalent bonds between N sp^2 - Fe and O - Fe. The adsorption energy of imidazoline is in the range of -1.7 eV and is higher than the adsorption energy of water. The SAM is composed of a network of imidazolines linked to each other through H-bonds. The influence of the alkyl chain length of imidazoline derivatives on the energetic and structural properties of the SAM was evaluated. Our results shows that the adsorption energy and cohesive energy of the SAM, as the tilt angle and the minimal distance between the alkyl chains, reaches a stable level when the number of carbon atoms of the alkyl chain is above 7-8.

[1] A. Gouron, K. Le Mapihan, S. Camperos, A. Al Farra, V. Lair, A. Ringuedé, M. Cassir, B. Diawara, New Insights in Self-Assembled Monolayer of Imidazolines on Iron Oxide Investigated by DFT, Corros. Sci. submitted

[2] J. Zhang, J. Liu, W. Yu, Y. Yan, L. You, L. Liu, Molecular modeling of the inhibition mechanism of 1-(2-aminoethyl)-2-alkyl-imidazoline, Corros. Sci. 52 (2010) 2059–2065.

Impact of ternary chloride salt melts on 9 % Cr steel

Jan MIDTLYNG[1]

[1] *Bundesanstalt für Materialforschung und -Prüfung (BAM), Germany*

Many chloride salts have high melting temperatures and have therefore been discussed as a storage material for high temperature heat storage. Ternary mixtures of chlorides offer lower melting temperatures and subsequently lower corrosion kinetic while still being able to store heat with high heat density.

In this work, the liquidus surfaces of ternary chloride systems were calculated and promising mixtures with melting temperatures below 300 °C were identified. P92 and iron alloyed with 9 % Chromium by weight were annealed in these low melting mixtures and the corrosion mechanisms were investigated by electron microprobe and by scanning electron microscopy.

The surface damage was observed applying 3D surface analysis.The results will be used in the development of a system of salt and container material with as low as possible corrosion.

Effect of cold-work on the oxidation resistance of AISI 304L in O2 atmosphere

Celine PASCAL[1], Audrey COL[1], Valerie PARRY[1]

[1] *Univ. Grenoble Alpes, CNRS, Grenoble INP, SIMaP, France*

In order to ensure a good resistance to high temperature oxidation, the oxide scales formed on stainless steels should remain thin, rich in Cr with good adhesion properties to the metallic substrate. However, the formation of the protective Cr-rich oxide scale induces a decrease of the Cr concentration in the bulk zone underneath the oxide scale. The balance between Cr consumption in the oxide scale and Cr diffusion from the bulk is a key parameter for the steel durability: below a critical level, Cr depletion leads to a loss of the scale protectiveness, and nucleation and fast growth of non-protective Fe-rich oxide nodules happens.

The Cr critical value needed to ensure a good oxidation resistance varies with the operating conditions or the steel metallographic state. High values are required for severe oxidation conditions such as wet or carburizing atmospheres. On the contrary, thinner alloy grain size or cold working before oxidation (grinding or traction) are believed to improve the steel durability. Since Cr diffuses faster along the grain boundaries than in the steel bulk, lowering the grain size improves the Cr effective diffusion. This effect is more pronounced in the fcc-structure of austenitic stainless steels where the lattice diffusion of Cr is slower than in bcc-structure of ferritic stainless steels. Moreover, a higher density of grain boundaries provides more oxide nucleation sites on surface, allowing efficient formation of a continuous oxide scale.

Mechanical treatments prior to oxidation may also improve the oxidation resistance. Since strain increases the concentration of defects (dislocations, grain boundaries), Cr diffusion is improved as well as the early formation of Cr-rich oxides.

This paper aims at studying the effects of cold working on the oxidation behaviour at high temperature of AISI 304L austenitic stainless steel. Specimens, subjected to 38 % strain using tensile testing at room temperature, were oxidized in O_2 at either 830 °C or 850 °C from 48 to 312 h. Evolution of grain size and surface hardness of the alloy substrate with oxidation time at 850 °C, show that the strain effects are withdrawn after 48 h of oxidation. Thermogravimetric analysis performed for 312 h at 830 °C indicate that the mass gain per surface unit area of the 38 % pre-strained sample is decreased by a factor two compared to the as-received sample. After 312 h at 830 °C, SEM investigations on sample surfaces and cross-sections show that the oxide scale formed on the pre-strained sample is thin while nodules locally appear on the as-received sample. After 48 h at 850 °C, Fe-rich oxide nodules appear on pre-strained sample but remains localized while for as-received samples a protective regime is never reached. In both cases, SEM investigations show that cold work promotes the formation of a protective duplex oxide scale from the first stages of oxidation that condition the long-term oxidation resistance.

Corrosion inhibition of copper, zinc and copper-zinc alloys using imidazole-based compounds with different functional groups

Peyman TAHERI[1], **Olmo BOTTAZZI**[1], **Ingrid MILOŠEV**[2], **Anton KOKALJ**[2], **Arjan MOL**[1]

[1] *Delft University of Technology, Netherlands*
[2] *Jožef Stefan Institute, Slovenia*

Corrosion inhibition of copper, zinc and copper-zinc alloys using imidazole-based compounds with different functional groups

Taheria, O. Bottazzia, I. Miloševb, A. Kokaljb, J.M.C. Mola

a Delft University of Technology, Department of Materials Science and Engineering, Mekelweg 2, 2628 CD Delft, The Netherlands.

b Jožef Stefan Institute, Department of Physical and Organic Chemistry, Ljubljana, Slovenia.

Cu and its alloys such as brass are used widely for different applications, e.g. electronics, instruments, pipelines, etc., due to their desirable mechanical and electrical properties. However, they are susceptible to corrosion and dealloying processes once exposed to corrosive media. Organic corrosion inhibitors are utilized to retard corrosion of the Cu alloys. A fundamental understanding of the interaction between corrosion inhibitors and metal surfaces is required to design sustainable systems, yet our knowledge about the corrosion inhibition mechanisms is still very limited. In this study, we evaluate interactions of imidazole-based compounds having different functional groups with a set of substrates, ranging from pure Cu to Cu-40%Zn alloy, and pure Zn. To this end, we utilize various experimental techniques. Moreover, the interfacial bonding of inhibitors to these substrates is also investigated by density functional theory (DFT) calculations. Polarization resistance (PR), electrochemical impedance spectroscopy (EIS) and Fourier transform infrared spectroscopy (FTIR) measurements show that the interactions of the imidazole-based inhibitors are relatively stronger on Cu than Zn surfaces, i.e., the coverage of inhibitors is larger on the former. Additionally, the Zn percentage plays an important role in the inhibition efficiency of the imidazole-based compounds where the surface crystallography defines the inhibitor-metal surface interaction mechanism. Moreover, the functional groups present in the inhibitor structure determines the interaction mode and consequently the inhibition mechanism. In compliance with experimental results, DFT calculations show that although imidazoles are active compounds, some addition functional groups, such as mercapto group, can significantly enhance the adsorption bonding and effectively retard corrosion of Cu and Cu alloys. Our findings indicate that both the metal surface properties and inhibitor structure influence the interfacial bonding mechanisms.

This work is a part of M.Era-Net project entitled "Coin Desc: Corrosion inhibiton and dealloying descriptors". The financial support of the project by NWO (Nederlandse Organisatie voor Wetenschappelijk Onderzoek) and MESS (Ministry of Education, Science and Sport of Republic of Slovenia) is acknowledged.

Keywords: Copper alloys; Imidazole-based inhibitors; DFT calculations; Alloying elements; Interfacial interactions; Electrochemical measurements.

Application of Advanced Fluorescence Techniques to Improve Scale Management in Conventional and Sub-Sea Deep Water Fields

Stephen HEATH[1], Susanna TOIVONEN[2], Neil POYNTON[1], Vesa VUORI[2], Hanna SUOMINEN[2]

[1] *Baker Hughes, a GE Company, United Kingdom*
[2] *Kemira, Finland*

Scale inhibitor analysis is an extremely important part of scale management and it is essential to have reliable methods for the accurate and precise measurement of scale inhibitor residuals in produced fluids in order to prevent wells from scaling and to make key decisions on the efficiency of scale squeeze and continuous chemical injection treatments especially in complex deep water environments with sub-sea wells.

In recent years the stricter environmental regulations in the North Sea, coupled with the development and operation of more complex fields in harsh scaling environments, has led to increased use of environmentally friendly polymeric scale inhibitors. The accurate and specific analysis of polymeric scale squeeze inhibitors is known to be difficult compared to phosphonates and has led to the development of a toolbox of advanced scale inhibitor analysis techniques including Fluorescence (F) and Time Resolved Fluorescence (TRF). These methods offer the potential to improve scale management capability in conventional and sub-sea fields through improved scale inhibitor detection at low levels.

Fluorescence techniques are particularly suited to tagged polymers and naturally fluorescent molecules whereas Time Resolved Fluorescence provides the ability to detect chemical functional groups, contained within polymeric scale inhibitor molecular structures, which were previously undistinguishable. The two techniques can be used individually or in combination and this provides a distinct advantage for multiple scale inhibitor analysis capability in produced brines where packages of different products can be selectively designed and detected for specific field applications. In addition, both techniques offer the capability of on-site detection compared to the majority of conventional analysis techniques which require onshore analysis.

The ability to detect polymeric scale inhibitors at very low MIC <1ppm with improved confidence has the potential for significantly extending scale squeeze lifetimes. In addition, this has now allowed highly efficient polymers to be used in field situations where scale squeezing had either been stopped or the lifetime was significantly compromised because of the lack of confidence in the scale inhibitor return profiles and the interferences from other scale inhibitors.

Specific examples from both sub-sea and conventional wells will be presented where the scale management has been or could be significantly improved through the application of the advanced fluorescence and TRF techniques. In addition, the two techniques will be compared and contrasted with each other and the pros, cons and limitations of both techniques will be presented.

MATERIAL CORROSION TROUBLES ON HEAT RECOVERY SECTION IN WASTE TO ENERGY FACILITY

Ha-Na JANG[1], Eun-song LEE[1]

[1] *Yonsei University, Korea, South*

Typically, municipal solid waste contains large portion of combustible matter to recover heat energy by thermal treatment. Conventional combustion is promising and well-known technology to treat municipal solid waste safely. However, municipal solid waste in Korea consists of miscellaneous materials with incombustible materials and large amounts of moisture, which prevent heat recovery from combusting. Due to fossil fuel limitation, technological development of renewable and sustainable energy such as waste to energy (WtE) is necessary. Solid refused fuel (SRF) mainly consists of combustible matter which is derived from municipal solid waste by separation process, and heat content was maximized by intensive drying process. Combustible matter in SRF contains plastics fraction to increase calorific value. In this study, commercial SRF boiler with 100MW capacity was selected to investigate corrosion issue in boiler tube and convection pass. Usually, chlorine content of fuel occur boiler tube corrosion, and alkali metals such as K, Na, Ca occur slagging and fouling. Based on the fuel analysis, the average chlorine content of SRF was 1.32%. In XRF analysis, the chlorine content in boiler tube ash and bottom ash was 6.1% and 4.3%, respectively. In SEM-EDX analysis, the chlorine content in corrosion part on boiler tube was ranged from 11.95% to 30.52%. Figure 1 shows the corrosion part of boiler tube and facility location which was occurred alkali-chlorine corrosion.

The formation of ceramic coatings on the surface of plasma-chemical reactor for biomass utilization

Zydrunas KAVALIAUSKAS[1], Dovilė GIMZAUSKAITE[1], Viktorija GRIGAITIENE[1], Liutauras MARCINAUSKAS[1], Vitas VALINCIUS[1], MIndaugas MILIESKA[1], Romualdas KEZELIS[1]

[1] *Lithuanian Energy Institute, Lithuania*

This paper reviews the plasma deposition method of alumina based ceramic coatings with regard to protect the moving parts of biomass boilers from chemical corrosion at high temperature. This problem is very important, when biomass with high content of heavy and alkali metal chlorides are processed. Chlorine containing fuels or waste tends to form corrosive deposits on the heat transfer surfaces of the boilers. In order to increase the lifetime of those surfaces the high quality corrosion-resistant alloy steels are used. However, it increase the production costs. Thus, there is a need to investigate the ways how to reduce the influence of corrosion of metallic surfaces. One of them is surface protection with material that is resistant to chemical corrosion at high temperature. The aluminum, zirconium, titanium oxide or its mixtures are the materials whose coatings possesses high thermal, corrosive and chemical resistance. Such coatings are used for surface protection against sulfur, chlorine, liquid sulfates and etc. The main aim of this research was to deposit the coatings using Al_2O_3 and mixture of Al_2O_3 with 3 wt.% TiO_2, 10 wt.% ZrO_2, $Al(OH)_3$, dolomite and zeolite powders on steel substrates by means of DC plasma torch at atmospheric pressure. The surface morphology of the as-sprayed and chemically treated coatings was characterized by scanning electron microscopy (SEM). The elemental composition of the coatings was investigated by energy dispersive X-ray spectroscopy (EDX). The coating structure was analyzed by X-ray diffraction (XRD). In order to determine the resistance to thermal treatment the deposited coatings were treated at temperatures of 350 °C and 500 °C. The duration of thermal treatment was one hour and 10 cycles were provided. The chemical resistance of the coatings was investigated using potassium chloride. The results indicated that the thermal treatment at 350 °C and 500 °C temperatures had no influence on the surface morphology and phase composition of coatings prepared from commercial and chromatographic Al_2O_3 powders. The XRD results indicated that the Al_2O_3 coatings consisted of alpha and gamma phases. The addition of TiO_2 or ZrO_2 increased the density and decreased the porosity of protective surface.

Acknowledgement: This research is funded by the European Regional Development Fund according to the supported activity 'Research Projects Implemented by World-class Researchers" under Measure No. 01.2.2-LMT-K-718.

Corrosion Inhibitors for Protection of Galvanized Steel in Aqueous Solutions

Vesna Alar[1], Ivan Stojanović[2], Boris A. Mikšić[3], Ivana Radić Boršić[4], John Wulterkens[5]

[1]*Faculty of Mechanical Engineering and Naval Architecture University of Zagreb, Croatia,* *vesna.alar@fsb.hr*
[2]*Faculty of Mechanical Engineering and Naval Architecture University of Zagreb, Croatia,* *ivan.stojanovic@fsb.hr*
[3]*FNACE, Cortec Corporation, St. Paul, Minnesota, USA, info@cortecvci.com*
[4]*Ecocortec d.o.o., Beli Manastir, Croatia, iborsic@cortecvci.com*
[5]*Cortec Corporation, St. Paul, Minnesota, USA, jwulterkens@cortecvci.com*

Abstract: This paper presents findings concerning the influence of corrosion inhibitors on the corrosion resistance of galvanized steel bolts in fresh water, 1% NaCl salt water, and distilled water. The inhibitors evaluated in this study were L-tryptophan, VpCI-377, Inhibitor 1240, and Inhibitor 7961. Quantitative tests performed include gravimetric analysis and electrochemical techniques, such as linear polarization, Tafel extrapolation, and Electrochemical Impedance Spectroscopy (EIS). Electrochemical measurements determined the polarization resistance R_p, corrosion rate v_{corr}, and corrosion potential E_{corr}. Furthermore, Scanning Electron Microscopy (SEM) and ATR-FTIR spectrometry were utilized to evaluate and characterize the formed layers on the surface of hot-dip galvanized steel samples. The potentiodynamic polarization curves show that all analysed inhibitors act as mixed type inhibitors protecting via adsorption on the metal surface. At 200 ppm, VpCI-377 showed the best protection of galvanized bolts in water.

Keywords: corrosion inhibitors, galvanized steel, aqueous solutions, corrosion testing

Introduction

Hot-dip galvanizing prevents corrosion by offering barrier and galvanic protection to the base steel. The galvanized coating is metallurgically bonded to the underlying steel, forming an impervious barrier between the steel substrate and the corrosive environment [1]. The use of zinc as a corrosion-resistant coating is usually limited to atmospheric corrosion, while in more aggressive environments it is necessary to protect steel with alternative corrosion protection methods. The corrosion rate of zinc in water and aqueous solutions can be significantly reduced by use of an appropriate corrosion inhibitor [2].

Sodium dichromate, borax, sodium silicate, and sodium hexametaphosphate show exceptional inhibition properties for zinc protection. Recently, the influence of organic, environmentally-friendly corrosion inhibitors for zinc protection is increasingly being investigated [3]. One of these environmentally-friendly corrosion inhibitors is L-tryptophan, an amino acid from the indole group, which showed good properties in zinc protection in sodium chloride solution at a concentration of 1×10^{-2} mol/L [4,5]. There is also an increasing use of adsorption inhibitors that create a thin monomolecular film on metal surfaces and prevent metal contact with aqueous solutions [6].

This paper presents findings concerning the influence of four corrosion inhibitors on the corrosion resistance of hot dipped galvanized steel bolts in fresh water, 1% NaCl salt water, and distilled water. The water is usually used immediately after the hot-dip galvanizing process for cooling the samples. The bolts are galvanized in perforated metal basket. Once the metallurgical reaction is complete and the fasteners are completely coated, the zinc is drained from the basket. The basket is then spun to remove excess zinc. The bolts are then cooled to room temperature by immersion in water. If the bolts are not properly dried before packing, the zinc corrosion (i.e. white rust) occurs.

Experimental study

The experimental study included gravimetric analysis and visual assessment, electrochemical polarization measurements, electrochemical impedance spectroscopy, SEM and EDX analysis and ATR-FTIR spectrometry. Electrochemical polarization measurements were used to determine the corrosion potential E_{corr}, polarization resistance R_p, and corrosion rate v_{corr}. Electrochemical Impedance Spectroscopy was conducted to determine the resistance of the inhibitor layer. Scanning Electron Microscopy (SEM) and EDX, as well as, ATR-FTIR spectrometry were utilized to evaluate and characterize the formed layers on the surface of hot-dip galvanized steel samples.

Materials and sample preparation

The study was conducted on galvanized steel bolts in fresh water, 1% NaCl salt water, and distilled water, with and without inhibitor. The evaluated inhibitors were L-tryptophan, VpCI 377, Inhibitor 1240, and Inhibitor 7961. All inhibitor were applied in the testing solution at a concentration of 200 ppm. Before corrosion testing, the thickness of the Zn-coating was measured using magnetic induction gauge was measured to be 40-50 μm.

VpCI 377 (sample 1.1, 1.2, 1.3) is a water-based, nitrite- and phosphate ester-free inhibitor designed as a complete replacement for oil-based preventives for indoor protection of equipment and components. It is a corrosion preventive liquid that meets tough antipollution requirements. VpCI-377 forms a solution in water and is applied to metal surfaces by spraying or dipping. It is not intended for application in chloride containing solutions.

Inhibitor 1240 (sample 2.1, 2.2, 2.3) dissolves and removes corrosion products and oxide deposits from the surface of the metal, cleans and conditions the surface for later processing. Effectively removes rust and residue from steel and galvanized steel, as well as from the most existing metals and alloys including aluminium. Fast-acting rust remover created on the metal surface creates a passivated layer, which is resistant to corrosion.

Inhibitor 7961 (sample 3.1, 3.2, 3.3) is designed as a complete substitute for corrosion protection with solvent-based coatings. Inhibitor 7961 is soluble in water and, with water, forms a transparent coating. It is applied on the metal surface by spraying or immersion. The protective coating is transparent.

L-tryptophan (sample 4.1, 4.2, 4.3) is an amino acid whose molecular structure consists of an indole ring, which is of particular importance in the field of research of environmentally acceptable corrosion inhibitors. In solutions containing nitrogen and oxygen, it creates a strong adsorption process on the metal surface, and protects the metal against corrosion.

Gravimetric testing and visual assessment

Gravimetric tests were performed by measuring the mass of the samples before (m_{before}) and after 24 hours (m_{after}) immersion in fresh water, 1% NaCl salt water, and distilled water, with and without inhibitor. The inhibitors were added to the test solutions at a concentration of 200 ppm. Before the immersion, the samples were degreased in ethanol and dried.

DC polarization measurements

The polarization measurements were carried out with a Potentiostat/Galvanostat Model 273 EG&E device in a standard three electrode test cell in relation to the reference saturated calomel electrode (SCE), with known interaction potential of +0,242 V versus a standard hydrogen electrode (SHE).

Polarization curves were registered after 600 seconds of exposure to 400 ml of testing solution in order to allow corrosion potential (E_{corr}) stabilization. Polarization resistance (R_{p}) was determined in a ± 20 mV potential range around the corrosion potential. During the potentiodynamic measurements, the working electrode was polarized to the potential of ± 250 mV relative to the corrosion potential and the current (I_{corr}) response was measured. Based on measured data unprotected and protected galvanized steel samples, the corrosion rate (v_{corr}) was calculated.

Electrochemical impedance spectroscopy

The protective properties of the inhibitor layer were investigated by electrochemical impedance spectroscopy [7], with a VersaSTAT 3 Potentiostat/Galvanostat EG&G PAR 378. The measurements were carried out after 24 h of immersion in the testing solution at room temperature (23 ± 2) °C, with and without inhibitor. The impedance spectra were performed at open circuit potential (OCP) with a 10 mV sinusoidal amplitude. The frequency range was from 100 kHz to 10 mHz. A three-electrode cell including a galvanized steel sample as the working electrode, a saturated calomel electrode (SCE) as the reference electrode and graphite sticks as the auxiliary electrodes were used in the experiments. The surface of the working electrode was 1 cm^2. The Solartron Z-View 2.2 software was used to interpret data.

SEM and EDX analysis and ATR-FTIR spectrometry

Scanning Electron Microscopy (SEM) and ATR-FTIR spectrometry were utilized to evaluate and characterize the formed layers on the surface of hot-dip galvanized steel samples. The surface coverage of the inhibitor layer was observed by a Tescan scanning electron microscope (SEM) equipped with Oxford Instruments energy dispersive spectroscopy (EDX). ATR-FTIR spectrometry was performed at PerkinElmer Spectrum One FTIR Spectrometer.

Results and discussion

Gravimetric analysis

From the results obtained after 24 hours immersion of the samples in aqueous solutions with and without inhibitors, it is apparent that the VpCI 377 inhibitor (samples 1.1 and 1.2) in water and distilled water showed best results, that is, no visible corrosion occurrence, whereas in 1% NaCl solution the surface changes are more pronounced. On samples protected by inhibitors 1240 and 7961, the damage was found in all testing solutions, while L-tryptophan resulted in white deposits on the surface of the samples. The results are shown in Table 1.

Table 1. Samples of galvanized bolts after 24 hours in solutions with and without inhibitors

Solution	No inhibitor	VpCI-377	1240	7961	L-tryptophan
Fresh water (samples 1)	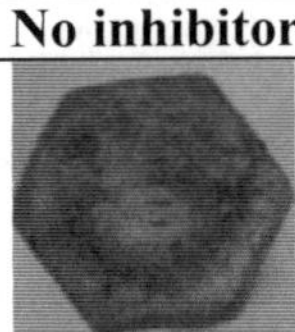	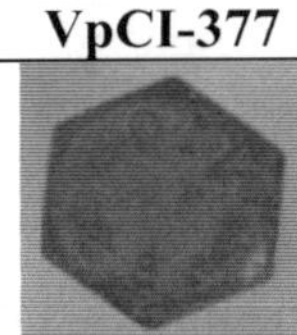			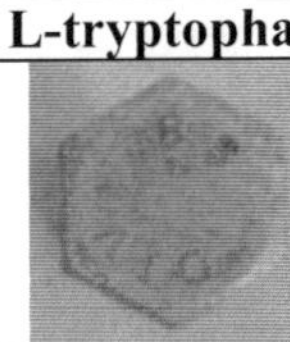
Distilled water (samples 2)		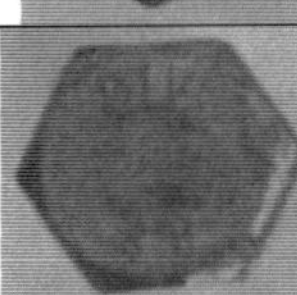	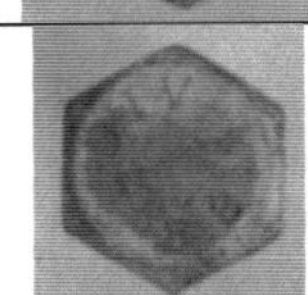	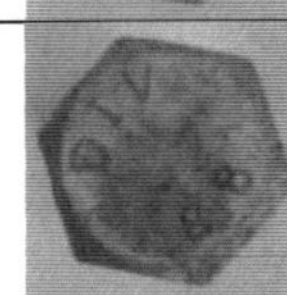	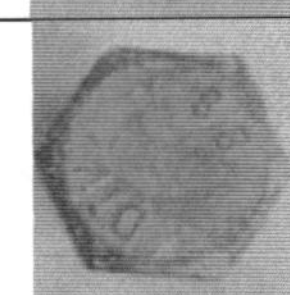
1% NaCl (samples 3)					

Gravimetric analysis showed that almost all samples lost the weight after 24 hours of immersion. The highest mass loss occurred on a sample immersed in a 1% NaCl solution treated with VpCI 377 (Sample 1.3). On the sample treated L-tryptophan in distilled water, mass growth was observed, resulting from white deposits on the surface of the screws. Table 2 shows the gravimetric test results before and after treatment in aqueous solutions.

Table 2. Gravimetric test results before and after treatment in aqueous solutions

Sample	m_{before} [g]	m_{after} [g]	Δm [g]
1	21,836	21,7580	0,078
2	21,5806	21,5313	0,0493
3	21,7563	21,5850	0,1713
1.1	21,8906	21,8901	0,0005
1.2	21,8549	21,7988	0,0561
1.3	21,8738	21,5759	0,2979
2.1	21,7975	21,7567	0,0408
2.2	21,7603	21,7583	0,002
2.3	21,7815	21,7749	0,0066
3.1	21,5766	21,5683	0,0083
3.2	21,5298	21,5131	0,0167
3.3	21,7823	21,7807	0,0016
4.1	21,8018	21,7993	0,0025
4.2	21,7393	21,7464	-0,0071
4.3	21,8077	21,8070	0,0007

Open circuit potential measurement (OCP)

Open circuit potential or corrosion potential (E_{corr}) was determined in relation to the reference saturated calomel electrode (SCE) on a surface of 1 cm^2 in 400 ml of water, distilled water and a 1% NaCl solution. The concentration of each inhibitor was 200 ppm, while 0,816 g of L-tryptophan was used. The results of corrosion potential measurements are given in Figure 1 to 3. The most negative values of corrosion potential were measured in 1% NaCl solution, with and without corrosion inhibitors. The results are presented in detail in Table 3, 4 and 5.

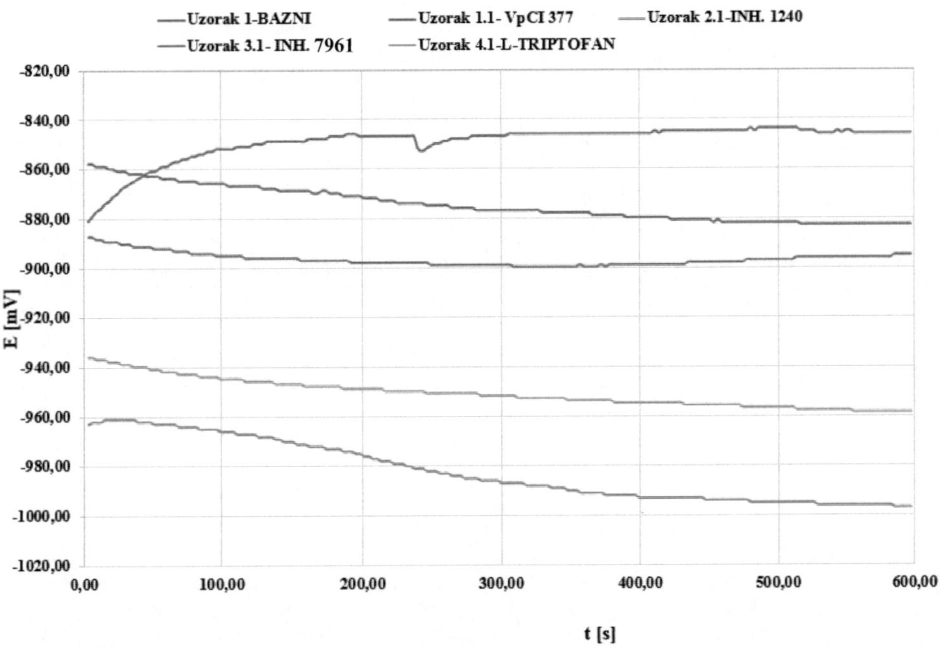

Figure 1. Corrosion potential diagram with and without the addition of inhibitors in fresh water

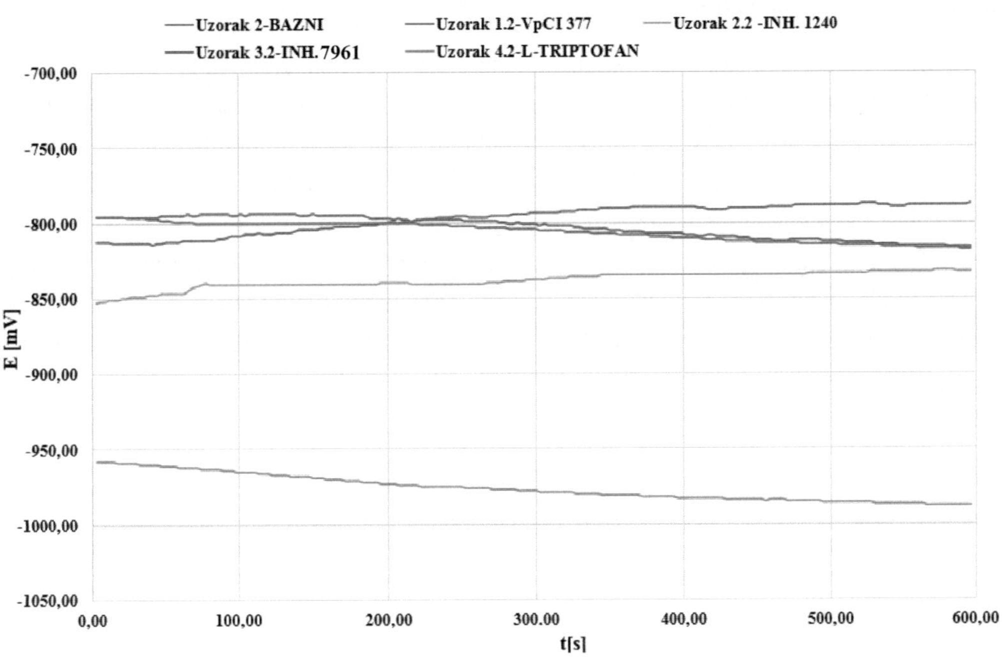

Figure 2. Corrosion potential diagram with and without the addition of inhibitors in distilled water

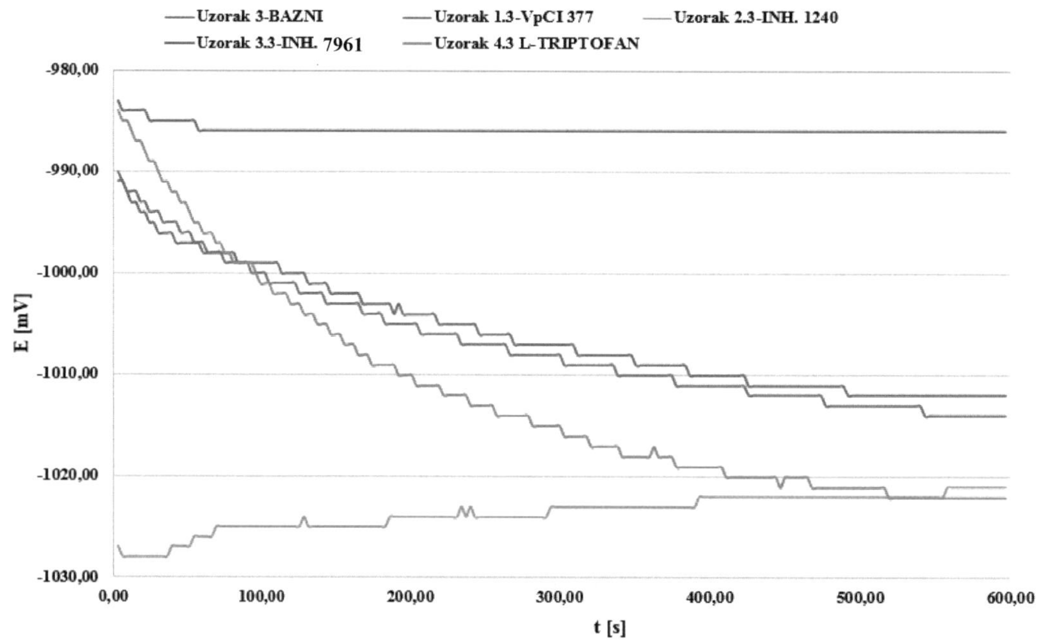

Figure 3. Corrosion potential diagram with and without the addition of inhibitors in 1% NaCl

Linear polarization measurements

The polarization resistance (R_p) is determined in the potential interval of ± 20 mV vs. E_{corr}. Higher polarization resistance means lower corrosion current and lower corrosion rate. The results of the polarization resistance tests showed that all analysed inhibitors provide good corrosion resistance in the fresh water. For the samples tested in distilled water, VpCI 377 showed the highest polarization resistance (R_p=60.55 kΩcm^2), while the worst resistance properties were displayed by Inhibitor 1240 and L-tryptophan. On the samples tested in 1% NaCl solution, the highest polarization resistance showed L-tryptophan (R_p=1.001 kΩcm^2), and the worst properties were achieved by Inhibitor 7961. Linear polarization results are shown in Table 3, 4 and 5.

Tafel extrapolation

Tafel extrapolation measurements provide important information about the kinetics of anodic and cathodic reactions [8]. Comparing the determined corrosion rate, it was found that VpCI 337 showed the best protection properties in fresh water and distilled water, while L-tryptophan shoed the best protecting in 1% NaCl solution.

The corrosion parameters, namely corrosion potential (E_{corr}), corrosion current density (j_{corr}), corrosion rate (v_{corr}), anodic and cathodic Tafel slopes (βa, βc) are listed in Table 3, 4 and 5.

Table 3. Polarization test results in fresh water, at a temperature (23 ± 2) °C, with and without inhibitor

Sample	*E* [mV]	R_p [kΩcm²]	j_{corr} [μA]	v_{corr} [mm/god]
1	-895	4,721	21,06	315,3 x 10^{-3}
1.1	-883	11,32	3,477	52,05 x 10^{-3}
2.1	-959	9,289	5,795	86,75 x 10^{-3}
3.1	-846	17,90	10,56	158,1 x 10^{-3}
4.1	-997	8,659	22,84	341,9 x 10^{-3}

Table 4. Polarization test results in distilled water, at a temperature (23 ± 2) °C, with and without inhibitor

Sample	*E* [mV]	R_p [kΩcm²]	j_{corr} [μA]	v_{corr} [mm/god]
2	-818	26,42	3,221	48,21 x 10^{-3}
1.2	-816	60,55	2,178	32,61 x 10^{-3}
2.2	-833	9,421	17,06	255,4 x 10^{-3}
3.2	-787	29,70	6,978	104,5 x 10^{-3}
4.2	-988	19,34	11,26	168,5 x 10^{-3}

Table 5. Polarization test results in 1% NaCl, at a temperature (23 ± 2) °C, with and without inhibitor

Sample	*E* [mV]	R_p [kΩcm²]	j_{corr} [μA]	v_{corr} [mm/god]
3	-1014	0,456	17,99	269,3 x 10^{-3}
1.3	-986	0,585	32,84	491,6 x 10^{-3}
2.3	-1021	0,984	13,31	199,3 x 10^{-3}
3.3	-1012	0,411	16,23	243 x 10^{-3}
4.3	-1022	1,001	12	179,6 x 10^{-3}

In fresh water, Inhibitor 7961 acted as an anodic inhibitor, while all other inhibitors showed cathodic inhibition behaviour. In distilled water, VpCI 377 and Inhibitor 7961 acted as mixed-type inhibitors; while using L-tryptophan and Inhibitor 1240, potential displacement to more negative values is observed. In 1% NaCl, the corrosion potential shift to more noble values is observed with application of VpCI 377 inhibitor indicating its anodic behaviour, while all other inhibitors showed cathodic behaviour.

The potentiodynamic polarization curves (Tafel diagrams) of uninhibited and inhibited galvanized steel samples after 1 hour in fresh water, distilled water and 1% NaCl solution are given in Figures 4, 5 and 6.

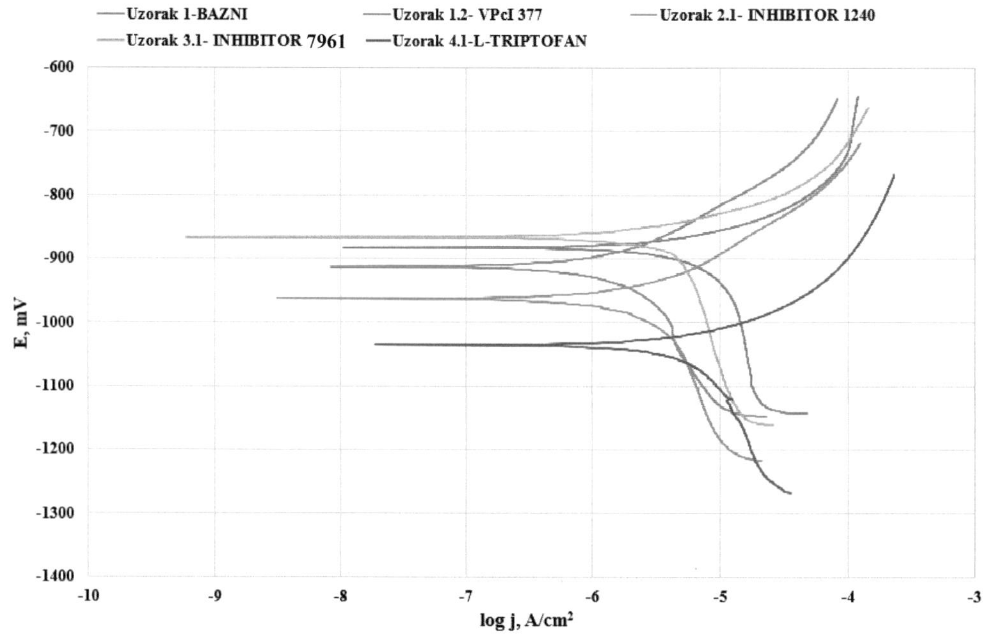

Figure 4. Polarization curves of tested inhibitors, compared to unprotected galvanized steel in fresh water

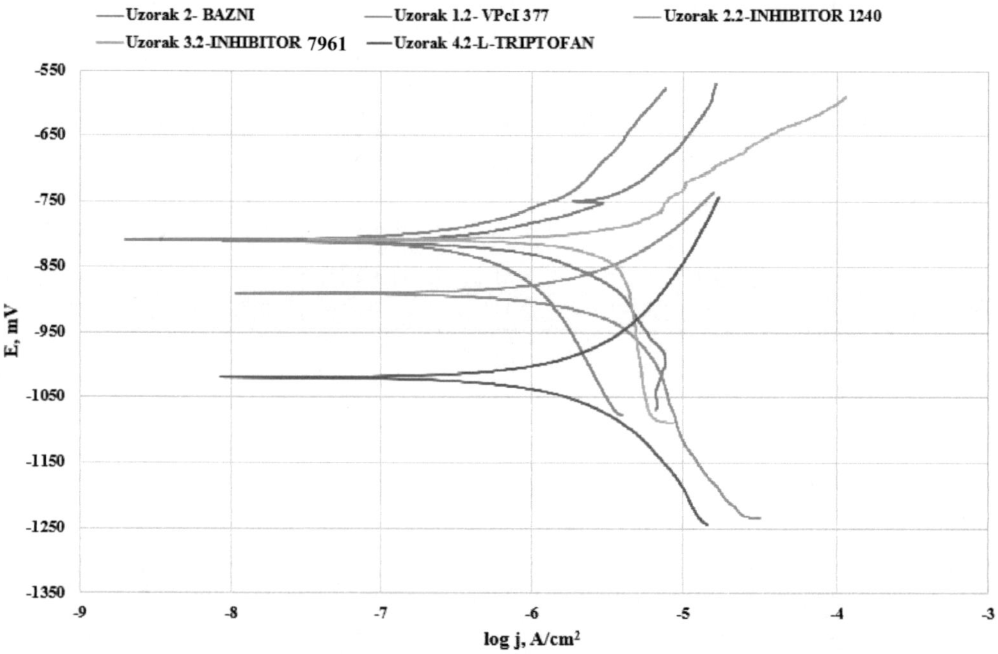

Figure 5. Polarization curves of tested inhibitors, compared to unprotected galvanized steel in distilled water

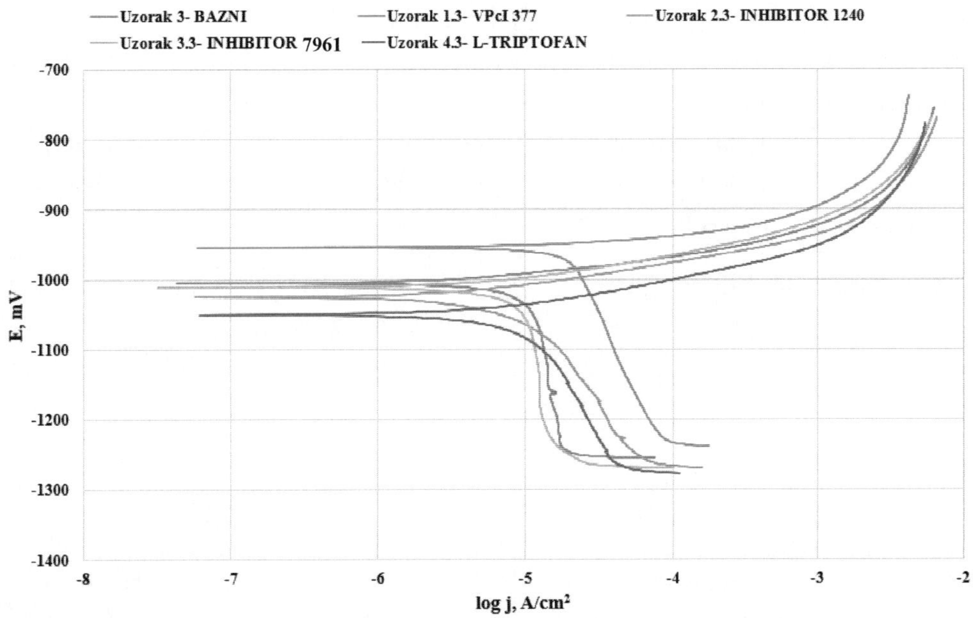

Figure 6. Polarization curves of tested inhibitors, compared to unprotected galvanized steel in 1 % NaCl solution

Electrochemical impedance spectroscopy (EIS)

After 24 hours of exposure to testing solutions, the impedance spectra was analysed by fitting with a suitable equivalent circuit model, given in Figure 7. The first model with Warburg element (W) describes the diffusion limited impedance by a layer (film, coating) undergoing corrosion at the metal surface. The second and third model the circuit elements shown are as follows: R_S = solution resistance, R_{coat} = coating resistance, R_{ct}= charge transfer resistance, C_{coat} = coating capacitance, and C_{dl} = coating double layer capacitance. The charge transfer resistance and double layer capacitance elements appear for a coating system with corrosion occurring on the metal surface [9].

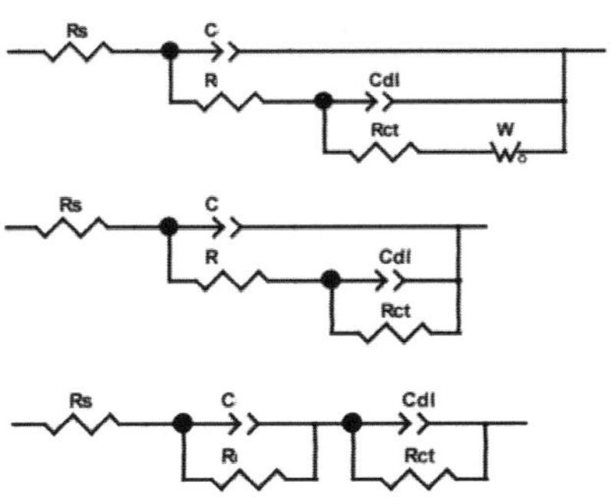

Figure 7. Electrical equivalent circuits [9]

The electrochemical behaviour at the coating/electrolyte interface is the key factor in understanding the protective properties of the inhibitor layer. After 24 hours, both in fresh water and 1% NaCl, L-tryptophan formed thick and hard removable deposit on the surface of the galvanized steel samples, acting as a barrier and thus resulting in highest layer resistance.

These white deposits of L-tryptophan changed the surface of galvanized steel and had a negative influenced on the functionality of fasteners, which is not allowed. The coverage of the thick surface layer of the L-tryptophan product also made obtaining the impedance spectra more difficult [10]. From other three tested inhibitors, VpCI 377 showed best protective properties (R_{sum}=3,40x10^5 Ωcm^2) in fresh water, while in 1% NaCl the best corrosion protection was delivered by Inhibitor 1240.

The EIS data, namely solution resistance (R_1), inhibitor layer resistance (R_2), charge transfer resistance (R_3), constant phase element of the inhibitor layer (CPE_1) and constant phase element of the double layer (CPE_2) with empirical constant n_1 and n_2, Warburg element (W) and total resistance (R_{sum}) are given in Table 6.

Table 6. EIS results after 24 hours immersions in fresh water and 1% NaCl, with and without inhibitor

Sample	R_1 [Ωcm^2]	R_2 [Ωcm^2]	R_3 [Ωcm^2]	CPE_1		CPE_2		W [Ωcm^2]	R_{sum} [Ωcm^2]
				Q_1 [F/cm^2]	n_1	Q_2 [F/cm^2]	n_2		
Zn in fresh water	2,55x10^2	1,83x10^3	9,48x10^2	9,27x10^{-7}	1	3,04x10^{-5}	0,75	/	3,03x10^3
Zn in 1%NaCl	8,78x10^1	2,45x10^2	2x10^2	2,65x10^{-5}	0	7,86x10^{-5}	0	3,39x10^{-3}	5,33x10^2
377 in fresh water	7,41x10^2	7,55x10^4	2,64x10^5	3,21x10^{-7}	1	2,26x10^{-6}	0,91	/	3,40x10^5
377 in 1%NaCl	2,89x10^1	7,65x10^2	1,29x10^3	1,66x10^{-5}	0,74	1,58x10^{-3}	0,61	/	2,08x10^3
7961 in fresh water	2,4x10^1	9,09x10^4	2,07x10^4	5x10^{-7}	1	2,38x10^{-5}	1	/	1,12x10^5
7961 in 1%NaCl	4,86	5,13x10^2	3,91x10^2	5,44x10^{-5}	0,61	6,33x10^{-3}	0,71	/	9,09x10^2
1240 in fresh water	1,51x10^2	3,82x10^4	3x10^4	1,82x10^{-7}	1	1,36x10^{-6}	1	/	6,84x10^4
1240 in 1%NaCl	2,03	1,96x10^3	7,47x10^2	1,07x10^{-5}	1	2,88x10^{-3}	0,72	/	2,71x10^3
L-trypt. in fresh water	1,13x10^1	9,61x10^7	2,19x10^8	6,15x10^{-9}	1	3,52x10^{-11}	1	6,61x10^{-15}	3,15x10^8
L-trypt. in 1%NaCl	1x10^2	8,41x10^6	3,56x10^7	2,84x10^{-10}	0,96	1,4x10^{-9}	0,93	4,61x10^{-8}	4,40x10^7

SEM and EDX analysis

SEM analysis showed irregularities on all examined samples i.e. an uneven layer of zinc coating on the surface or chloride depositions from the aqueous solutions. On the samples treated with L-tryptophan, sedimentation of white clusters was visible on the surface. EDX analysis of the tested samples in aqueous solutions with and without inhibitors revealed the chemical composition of adsorbed layers on the galvanized steel samples. L-tryptophan treated samples were found to contain oxygen, which is very important for the process of adsorption of the inhibitor compound on the metal surface. The presence of oxygen in EDX analysis was also confirmed on samples protected with vapour phase corrosion inhibitor VpCI 377, showing that both corrosion inhibitors act as an adsorption inhibitor. EDX of Inhibitor 7961 showed chloride content in surface layer which can cause porosity in the protective film.

VpCI 377

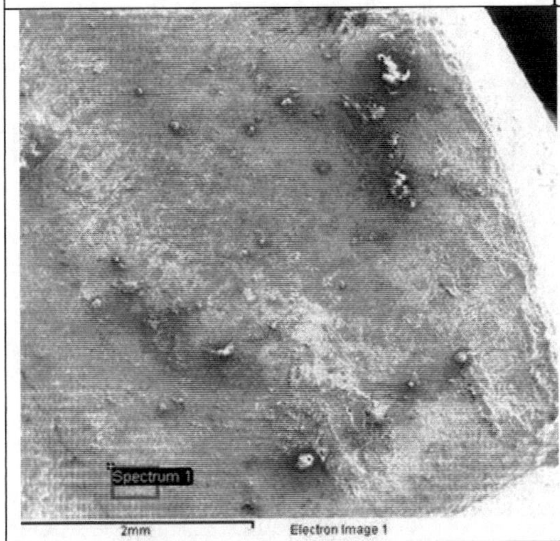

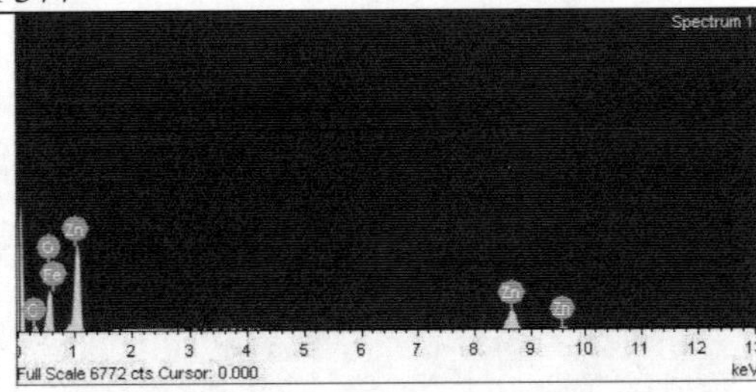

Element	Weight%	Atomic%
C	13,01	26,5
O	35,37	54,11
Fe	0,91	0,4
Zn	50,71	18,99

INHIBITOR 7961 (uzorak 3)

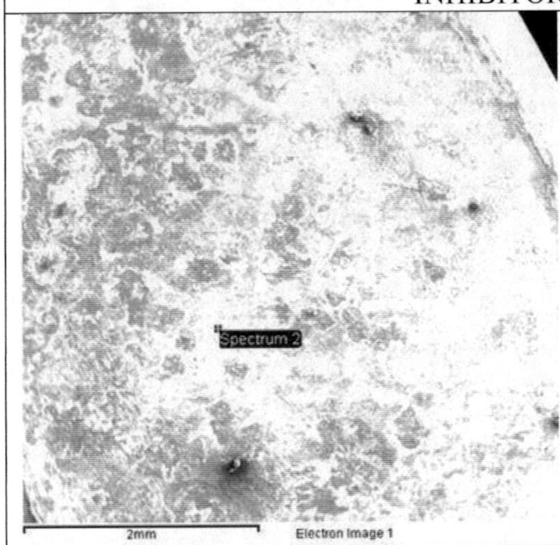

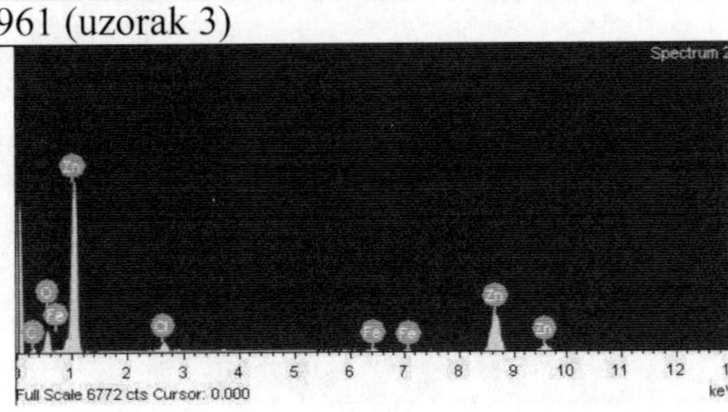

Element	Weight%	Atomic%
C	10,68	29,75
O	14,83	31,03
Cl	2,16	2,04
Fe	1,84	1,10
Zn	70,49	36,08

L-TRIPTOFAN

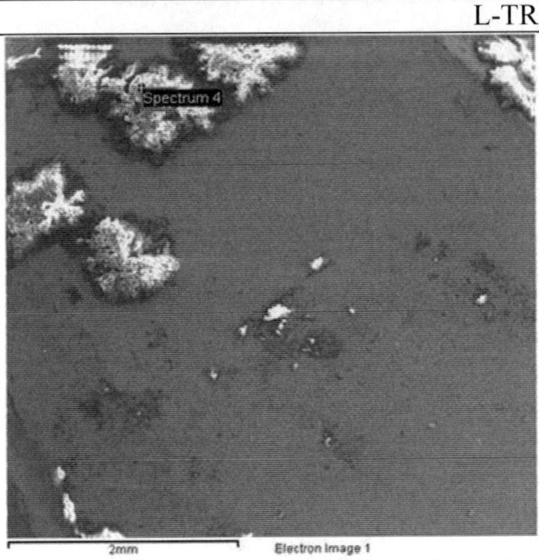

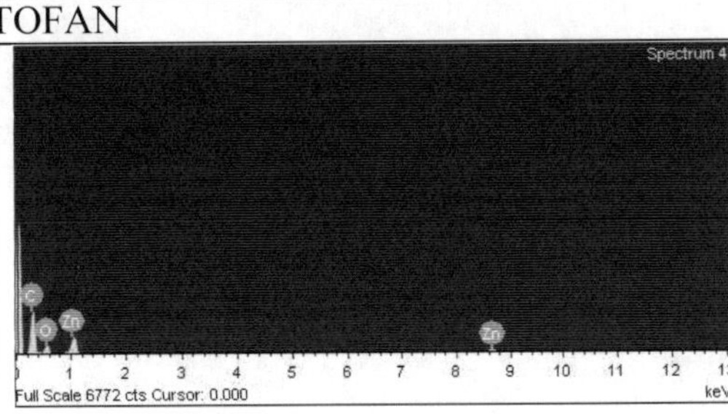

Element	Weight%	Atomic%
C	65,2	76,98
O	23,11	20,48
Zn	11,69	2,54

Figure 8. SEM view of inhibitor layer after 24 hours in fresh water

VpCI 377

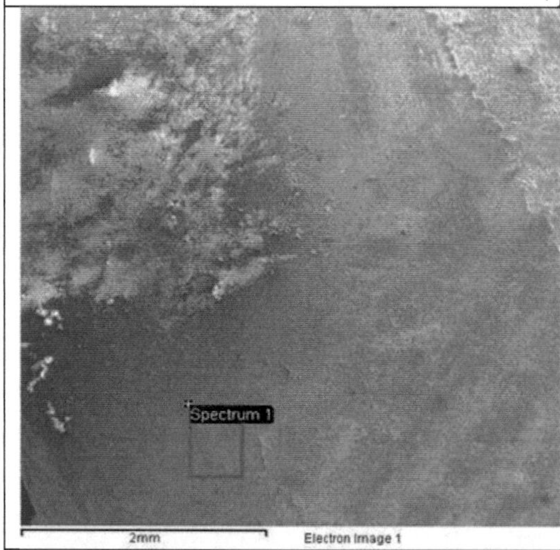

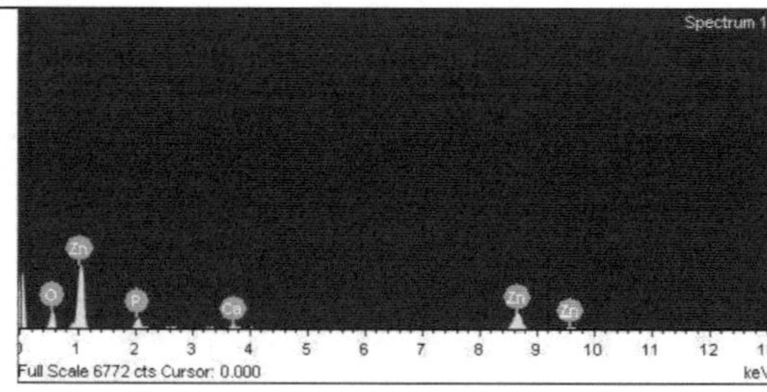

Element	Weight%	Atomic%
O	23,59	53,06
P	6,65	7,72
Ca	2,32	2,08
Zn	67,45	37,13

INHIBITOR 7961

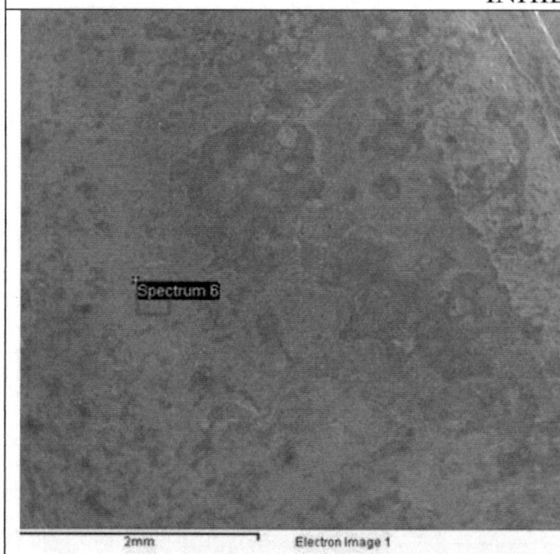

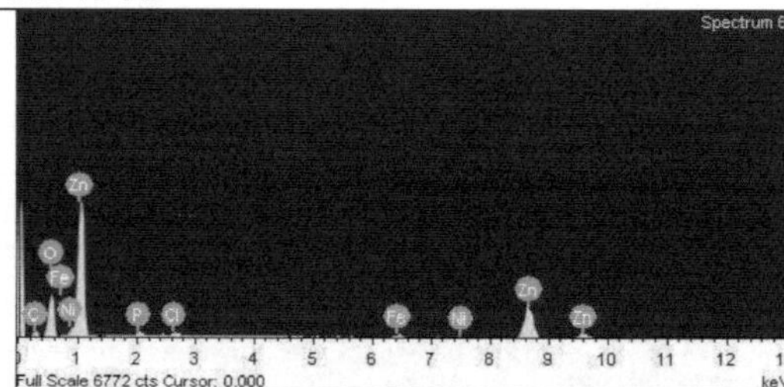

Element	Weight%	Atomic%
C	12,3	27,78
O	27,33	46,32
P	1,01	0,89
Cl	0,70	0,54
Fe	1,59	0,77
Ni	0,66	0,30
Zn	56,4	23,4

L-TRIPTOFAN

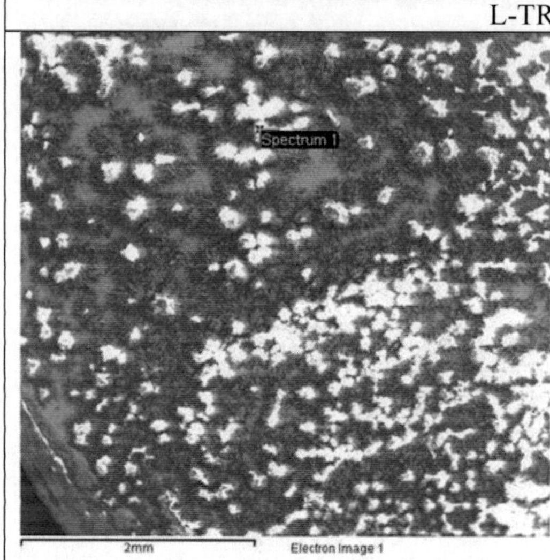

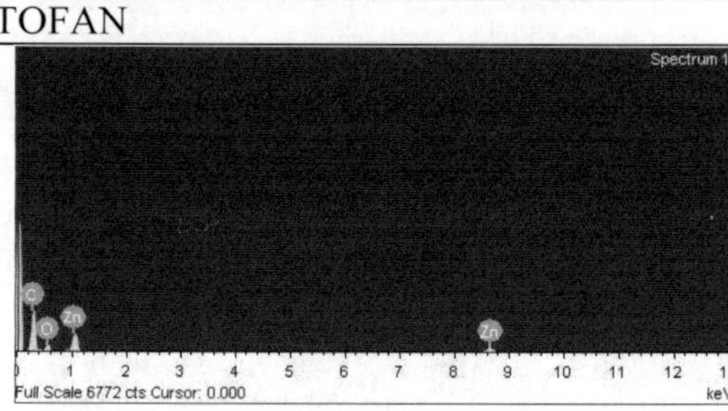

Element	Weight%	Atomic%
C	64,03	79,67
O	17,16	16,03
Zn	18,81	4,3

Figure 9. SEM view of inhibitor layer after 24 hours in distilled water

ATR-FTIR spectrometry

FTIR spectra of tested inhibitors on galvanized steel samples in fresh water are given in Figure 10.

The IR spectra of VpCI 377 showed several strong and strong to medium peaks which belong to carboxylic acid. The C=O stretch appears at 1504 cm^{-1} and the C-O stretch at 1390 cm^{-1}. The 163 cm^{-1} difference between asymmetric and symmetric stretching vibrations corresponds to bidentate binding of carboxylate group to Zn. The observed low intensity vibrations at 471 cm^{-1} correspond to Zn-O stretching vibration.

The IR spectra of Inhibitor 1240 showed several strong and strong to medium peaks which belong to carboxylic acid. The C=O stretch appears at 1103 cm^{-1} and the C-O stretch at 1034 cm^{-1}. The 69 cm^{-1} difference between asymmetric and symmetric carboxylate stretching vibrations corresponds to unidentate binding of carboxylate group to Zn. The low intensity vibrations at 574 cm^{-1} correspond to Zn-O stretching vibration.

The IR spectra of Inhibitor 7961 did not show the exact type of bonding of the inhibitor to Zn. The bonding type in these acids (long-chain fatty acid,) is not ordinary (uni- or bidentate) thus it is not easy to describe by IR spectra. Absorption of the inhibitor to Zn could be achieved by several different mechanisms. One of the possibilities is that the bonding is formed in unoccupied spaces of acid chains.

The IR spectra of L-tryptophan showed the strong and broad absorption band in the range of 3600 to 3000 cm^{-1} which corresponds to asymmetric and symmetric stretching vibration of aqua molecules. Bands in the range of 3392 cm^{-1} and 3161 cm^{-1} belong to the N-H stretches of NH_2 group of tryptophan. The week bands in the 2936-2906 cm^{-1} range are attributed to the CH_2 vibrations [11]. In complexes, tryptophan ligands are coordinated to the metal ion as unidentate by carboxylic group. The $(COO^-)_{asym}$ peak is located at 1621 cm^{-1} for Zn complexes. The $(COO^-)_{sym}$ peak is observed at 1408 for Zn complexes. The low intensity bands in the region of 600-400 cm^{-1} are attributed to Zn-N and Zn-O vibration.

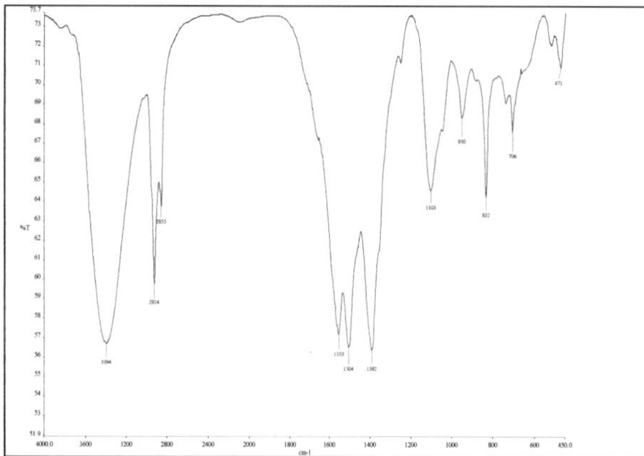

VpCI 377

Maximum cm^{-1}	Functional group	
471	Zn-O	stretch
832	C-H	stretch
1103	C-O	stretch
1392	C-O	stretch
1504	C=O	stretch
2924	C-H	stretch
3394	O-H	stretch

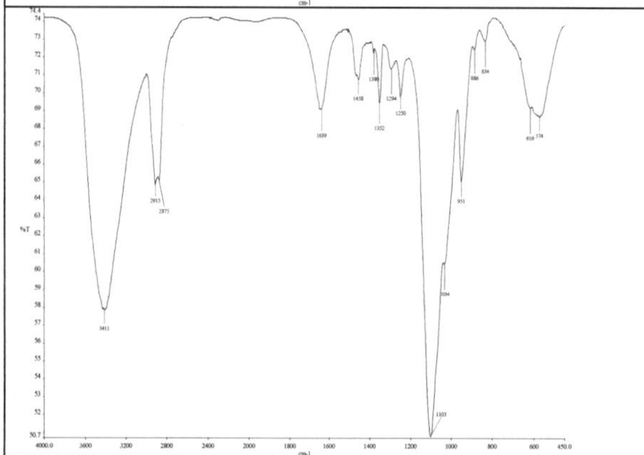

Inhibitor 1240

Maximum cm^{-1}	Functional group	
574	Zn-O	stretch
619	C-H	stretch
1034	C-O	stretch
1103	C=O	stretch
1639	C=O	stretch
2915	C-H	stretch
3411	O-H	stretch

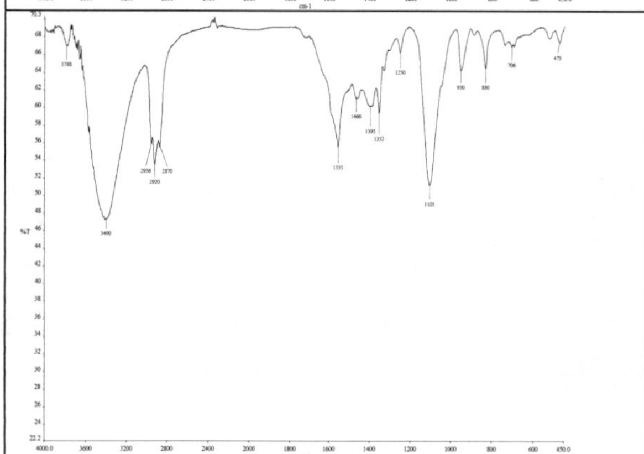

Inhibitor 7961

Maximum cm^{-1}	Functional group	
473	Zn-O	stretch
830	C-H	stretch
950	C-O	stretch
1105	C=O	stretch
1555	C=O	stretch
2870	C-H	stretch
2920	C-H	stretch
2956	O-H	stretch
3400	O-H	stretch

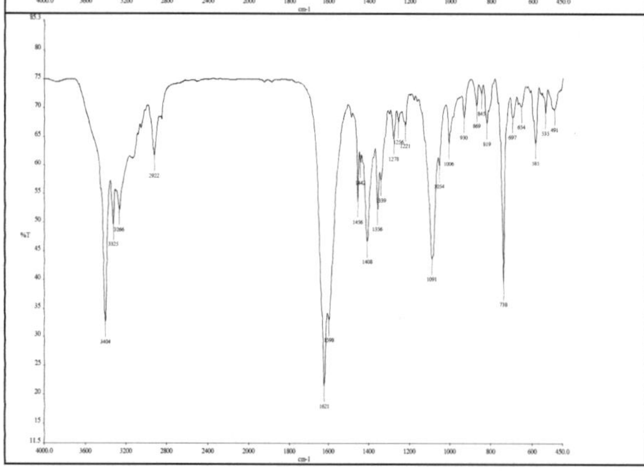

L-tryptophan

Maximum cm^{-1}	Functional group	
491	Zn-N	stretch
585	Zn-O	stretch
738	C-N	stretch
1091	C-O	stretch
1408	C-O	stretch
1598	N-H	bend
1621	C=H	stretch
2922	C-H	stretch
3404	N-H	stretch

Figure 10. FTIR spectra of tested inhibitors on galvanized steel samples in fresh water

Conclusion

The results of this study are summarized as follows:

- The results of the coating thickness on the galvanized bolts samples show that the arithmetic mean of some samples is below the permissible limit of 50 μm.
- Gravimetric testing determined the mass loss of almost all samples after 24 hours immersion in aqueous solutions with and without inhibitors: Only the sample treated with L-tryptophan in distilled water showed a mass increase, resulting from the formation of white deposits on the sample surface.
- Electrochemical DC tests showed that the best corrosion resistance in fresh water and distilled water was achieved with VpCI 377 inhibitor, while L-tryptophan showed best results in 1% NaCl solution.
- Electrochemical impedance spectroscopy showed that L-tryptophan provide highest layer resistance, but negatively influenced on surface functionality of the fasteners and their aesthetic requirements. VpCI 377 showed best protective properties in fresh water and distilled water without changing surface condition of the bolts.
- SEM and EDX analysis of galvanized steel samples showed higher oxygen content in L-tryptophan inhibitor film, which is very important for the adsorption process and the formation of a protective inhibitor layer on the zinc coating.
- FTIR spectra of tested inhibitors on galvanized steel samples showed that bonds of inhibitors VpCI 377, Inhibitor 1240 and Inhibitor 7961 belong to a carboxylic acid group, whereas L-tryptophan to amino acids.

References

[1] Hot-Dip Galvanized Fasteners: Advantages of and design considerations for hot-dip galvanized fasteners, American Galvanizers Association, Colorado, 2009.

[2] I. Juraga, V. Alar, I. Stojanović: Korozija i zaštita premazima, Zagreb, FSB, 2014.

[3] F.L. Laque, H.R. Copson, Corrosion Resistance of Metals and Alloys, Reinhold Publishing Corporation/Chapman & Hall Ltd, New York, London, 1963.

[4] Jia-Jun Fu, Su-Ning Li, Lin-Hua Cao, Ying Wang, Lian-He Yan, Lu-De Lu: L-Tryptophan as green corrosion inhibitor for low carbon steel in hydrochloric acid solution, J Mater Sci (2010) 45:979–986

[5] N. Wint, J. H. Sullivan, and D. J. Penney: The Role of pH on the Inhibition of Aqueous Zinc Corrosion by L-tryptophan, Journal of The Electrochemical Society, 164 (7) C356- C366 (2017)

[6] Boris A. Mikšić, FNACE: VpCI Technology Handbook, vol. 1, USA, 2014.

[7] Bierwagen, G.; Tallman, D.; Li, J.; He, L.; Jeffcoate, C. EIS studies of coated metals in accelerated exposure. Prog. Org. Coat. 2003, 46, 148–157.

[8] A. Ehsani, H. Mohammad Shiri, M. G. Mahjani, R. Moshrefi, R. Safari, Kovove Mater, 54 (2016) 233–239.

[9] U.S. Department of the Interior Bureau of Reclamation Reasearch and Development Office, Coating Evaluation by Electrochemical Impedance Spectroscopy (EIS), December 2015.

[10] A. Prasad Yadav: Electrochemical Impedance Response of Zn and Galvanized Steel Corroding under Marine Atmospheric Environments, J. Nepal Chem. Soc., Vol. 23, 2008/2009, 33-42.
[11] A. Kaşarcı, D. Ali Köse1, G.Alp Avcı, E. Avcı: Synthesis and Characterization of Co^{II}, Ni^{II}, Cu^{II} and Zn^{II} Cation Complexes with Tyrptophan. Investigation of Their Biological Properties, Hacettepe J. Biol. & Chem., 2013, 41 (2), 167–177.

Effect of the Cerium (III) chloride on the corrosion inhibition of aluminium alloy

Marin KURTELA[1], Vinko ŠIMUNOVIĆ[1], Vesna ALAR[1], Suzana JAKOVLJEVIĆ[1]

[1] *Faculty of Mechanical Engineering and Naval Architecture, Croatia*

The corrosion protection of aluminium alloy AA2024-T3 has been studied in NaCl with and without the addition of Ce (III) chloride. Various different exposure periods of immersion into 0.01 M NaCl, containing 0.003 M of $CeCl_3$ x H_20, are investigated to determine the effects of the inhibitor on the metal surface. Quantitative tests performed include electrochemical techniques, such as method of quasi - potentiostatic polarization (Tafel extrapolation) to determine corrosion rate v_{corr}, linear polarization to determine polarisation resistance R_p, inhibition efficiency (η %), protective ability (γ %) and electrochemical impedance spectroscopy (EIS). The samples were analyzed with SEM observation and EDX to evaluate and characterize the formed precipitates on the surface of aluminium samples. Cerium (III) chloride has been tested to be an effective corrosion inhibitor for AA 2024-T3 in 0.01 M NaCl solution. The formation of precipitates of cerium chloride occurred on cathodic intermetallic sites and the surface in general resulted in improved corrosion resistance.

Keywords: Aluminium alloy AA2024-T3, corrosion inhibitors, corrosion testing.

Development of a Novel Phosphonate Scale Inhibitor for Scale Control in Various Harsh Environments

Bethanni MCCABE[1], **Steve HEATH**[1], **Joseph LEDSHAM**[2]

[1] *Baker Hughes, A GE Company, United Kingdom*

[2] *Baker Hughes, A GE Comapny, United Kingdom*

The challenges of scale inhibition faced by the oil industry are continuously changing as well as the chemical requirements associated with scale inhibition. When selecting a scale inhibitor there are a number of factors and limitations that will influence the chemistry chosen for each application and often there is not one chemical that is fit for purpose for scale control under both conventional and high temperature conditions. A novel phosphonate chemistry has been developed which bridges the performance gap between different phosphonate products to be applicable over a wide range of scaling conditions. In addition, this product can also be easily be blended with other chemicals like H_2S scavengers.

This study is focussed on the development of a multifunctional, biodegradable, calcium tolerant and thermally stable scale inhibitor that can be used for a variety of applications including topside through continuous injection, downhole via capillary application and also scale squeeze treatments.

The new phosphonate scale inhibitor has been designed to work for the three most common types of scale; calcium carbonate, calcium sulphate and mild to moderate barium sulphate and examples of these will be provided. Furthermore, for calcium carbonate, details will be provided of scale inhibition performance testing in a geothermal brine at 250°C, where the new phosphonate chemistry demonstrated significant improvements compared to other phosphonates and polymeric scale inhibitors known to be stable at high temperature.

Field examples will be also discussed including data from a high calcium brine in West Africa (30,000mg/L), where the chemical has been deployed via downhole capillary injection and for continuous topside injection in a North Sea field where this chemistry was used in combination with H_2S scavenger.

This paper discusses the development of a novel, biodegradable phosphonate scale inhibitor and highlights the potential multiple uses of this chemistry in a variety of challenging environments.

Characteristics and Formation-Growth Model of Corrosion Scales in Oil Wellbore

Yan YANG[1], Abhijeet YADAV[2], Rajan AMBAT[3]

[1]*DTU Mekanik, Building 425, 2800 Lyngby Denmark, yyang@mek.dtu.dk*
[2]*DTU Mekanik, Building 425, 2800 Lyngby Denmark, abyadav@mek.dtu.dk*
[3]*DTU Mekanik, Building 425, 2800 Lyngby Denmark, ram@mek.dtu.dk*

Abstract

Scaling formation in the wellbore is a serious problem and affects the corrosion behaviour of the pipeline materials by changing the morphology and the physiochemical properties of the surface layer. This paper presents the results of a comprehensive study of scale found in oil well bore used in North Sea oilfield. Scales were removed from the inside of pipes which were taken from different depth under sea level and investigated by several analytical methods including visual observation and digital photography of the interior surface of the pipe and scales, high resolution scanning electron microscopy, X-ray diffraction, Energy Dispersive X-Ray Spectroscopy etc. The characteristic images of scale were obtained by microscopic analysis and results showed scale with characteristic structural features including a dense layer close to the corroded steel surface, a porous layer of iron hydroxide and chloride-rich products, bulk layer of akagneite, a thin layer of copper sulphide and a granular layer. A conceptual model was developed in this paper to describe the formation and growth of scale based on the history of the well. According to the developed model, corrosion scales were expected to grow from inside the scale via the corrosion reaction. Scale structure and composition played important roles in the migration of chloride ions and the reactions of corrosion scales that lead to iron release.

Keywords: Scale deposit; Corrosion products; Morphology; Layered structure; Ion migration

Introduction

Scale and corrosion in the wellbore have a serious problem that always an issue for oilfield operators. The issue is more prominent for oil fields with high pressure or temperature, and high salinity and high scaling water, the scaling problem is more prominent [1-3]. Localized corrosion or under deposit corrosion attack usually occurs due to porous nature of the scale or once the protective scale is damaged locally, and constitutes a principle cause resulting in the failure of the pipelines [4-6]. Scaling also can reduce the cross-sectional area of the formation channel, flow area and the equipment operation ability, leads to wellbore corrosion perforation or even out of use, seriously affecting the normal production and the oil and gas production safety [7-10]. The most common scales encountered in oil field operations are sulphates such as calcium sulphate (anhydrite, gypsum), barium sulphate, strontium sulphate and carbonates. Other common scales have been reported such as iron oxides, iron hydroxides, iron sulphides and iron carbonate. These scale types are most commonly associated with iron generation from corrosion products [11-15]. Numerous studies on scaling prediction have been published over the past few years [16-17], however, scale prediction software is still not fully developed that can accurately predict the formation of scales. This is because the understanding of the scaling in the wellbore is not deep enough. Therefore, research on the formation and growth process and mechanism of scaling should be strengthened.

This paper focuses on the detailed systematic investigation of the scales and corroded internal surface of the pipes with corrosion products to understand the characterization of scales formed in the different depth of oil produced well. It has a good guiding significance for the improvement of the scaling prediction model and the settlement and protection of scaling problems.

Materials and preparation

Pipes were retrieved from the oil produced well located in the North Sea, to investigate the nature of the pipe corrosion by scales. Four section pipes marked #1 to # 4 were obtained from the top (200ft), upper (3300ft), middle (4600ft) and lower part (6000ft) of oil produced wellbore and the material for the pipe is L80 steel (Table 1). All of the pipes were approximately 1m long with an outside diameter of 4.5 inches. From each pipe, small specimens (the size is 1*1cm) were cut off for microscopic observation. The scale on the inner surface of the pipe is very fragile; it was easy to fall off during the cutting process. Therefore, the SEM sample was divided into two parts, one was the stripped scale, and another one was the corroded pipe with scale. Both parts of the samples were mounted with epoxy resin as shown in figure 1. Before experiments, each sample was polished with 800# to 4000# grit silicon carbide paper, and was finely polished with 3 μm and 1 μm diamond grains, cleaned with ethanol, dried with hot air and placed in a drying box.

Table 1 Chemical composition of L80 steel specimen (% wt)

C	Mn	Cr	Ni	Cu	P	S	SI	Fe
0.43	1.9	1	0.25	0.35	0.03	0.03	0.45	balance

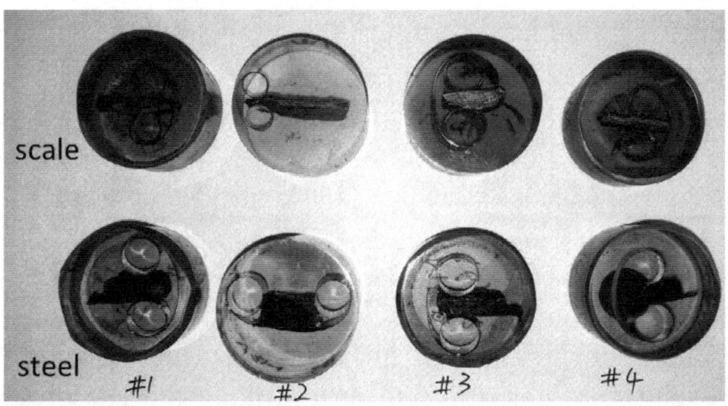

Fig.1 Samples of scale and corroded steel for SEM observation.

Analytical Methods

Scales were investigated by several analytical methods including visual observation and digital photography of the interior surface of the pipe, cross-section observation by high resolution scanning electron microscopy (SEM) of Quanta FEG 200 with Energy Dispersive X-Ray Spectroscopy (EDS) of Oxford Instruments X-Max Silicon Drift Detector. X-ray diffraction (XRD) is the most frequently applied method of testing the composition and concentration of crystalline phases in the corrosive products. Samples of scale, brought into the form of powder, were tested by diffractometer Bruker D8 under the following conditions: a Cu Kα radiation at 40 mA and 40 KV. The 2θ ranged from 10° to 120°, with a 0.02° step, and a 4 s count time at each step.

Scale structure and composition

Surface observation

After longitudinal sectioning of representative samples from each pipe, the inside surface was examined. Digital images were obtained prior to sampling as shown in figure 2. By visual examination, the scale inside the pipe is thick (4 to 10 mm) and the pipe was highly corroded. The pipe from top of the well (#1) showed the high amount of penetration and localized attack (as shown by black arrows) compared to other pipes from higher depth. The surface of the scale was a layer of yellow or brown material that was not continuous. Below the yellow layer was a black layer of corrosion product. The scale surface was uneven, some small tubercles on it. The colour of the scale was varied, indicating that it contains different phases.

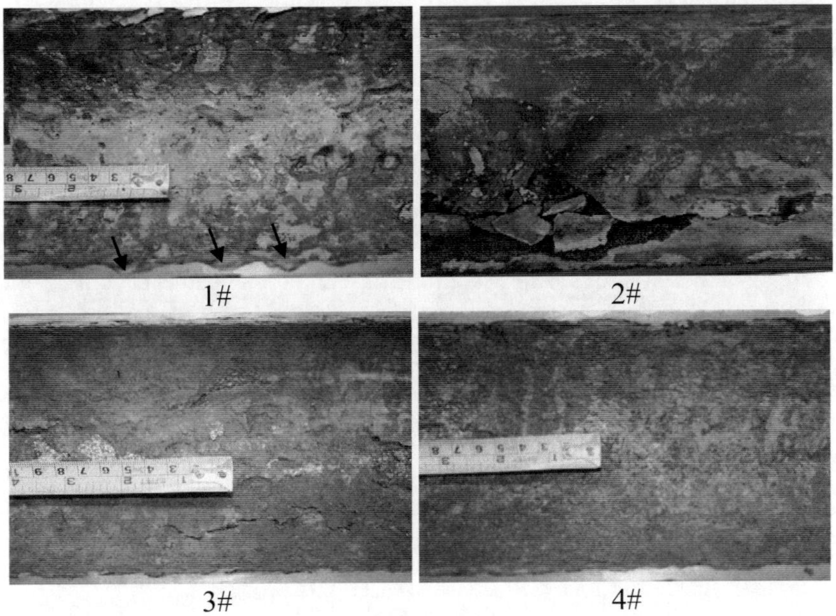

Fig.2 Scale surface of scale inside the pipes under different depths.

<u>SEM observation and Component analysis</u>

The scales formed on the L80 steel samples were analysed first. Figure 3 and 4 shows the cross-section of scale samples and corroded steel by products in different depths. The results showed that the scale cross-section of different depth has the similar structure. In order to fully understand the composition of scale, the scales were divided into 4 or 5 parts (named No.1 to 5) for EDS elemental analysis. On the oil side, there was a granule layer (marked No.1), which contained many irregularly shaped particles. The EDS results showed that the major constituents of the granule layer were barium, sulphur and oxygen, while some strontium and calcium was also found. By comparing the mapping images, it showed that the layer was mainly composed of barium, strontium and calcium sulphate (Fig.5 and 6). In addition, this layer was also doped with some impurities such as silica and some iron oxides or hydroxides. The bulk layer (marked No.3) was homogeneous with some cracks, which was very rich in iron, oxygen and chlorine, thus indicating presumably iron oxy-hydroxides. In the lower part of the bulk layer, the scale structure became loose, and some pore holes or fibrous structures appeared (marked No.4), which become more obvious with the increase of pipe depth, as shown in Fig.3 #4. Near the surface of the steel, there was a dense layer with some cracks (marked No.5). The boundary between steel and scale indicated that the corrosion of steel surface was serious.

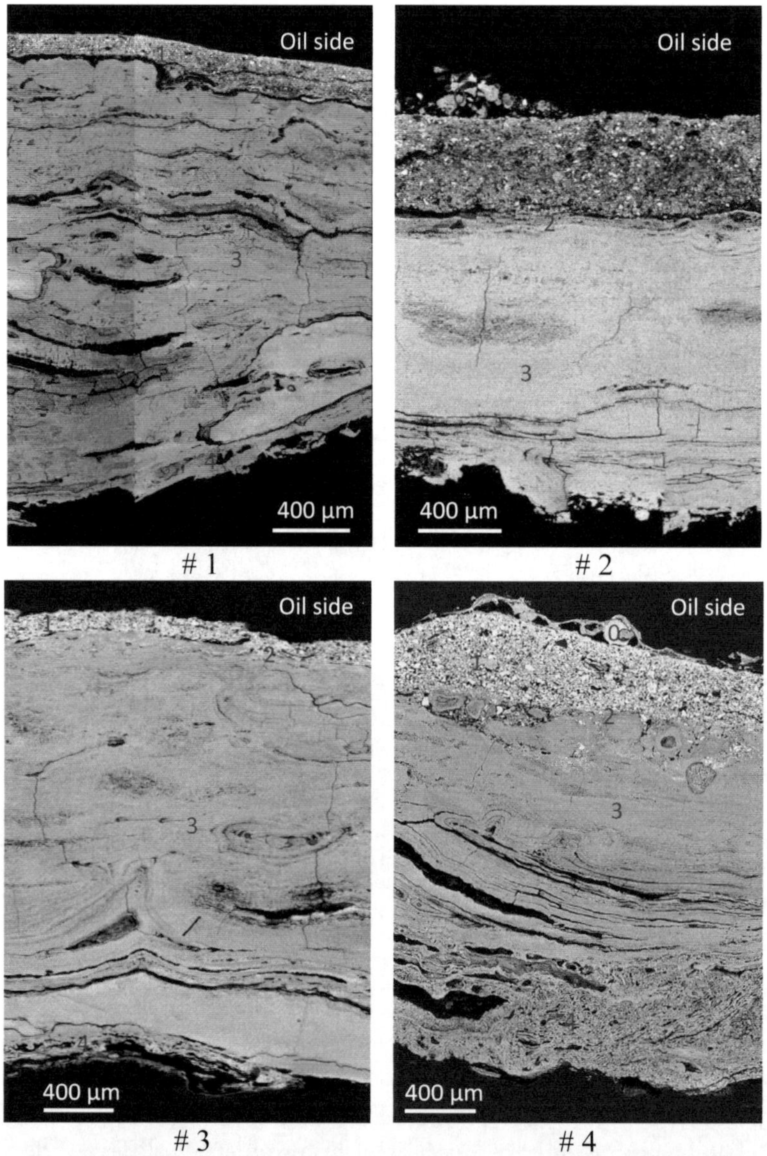

Fig.3 The cross section images of the scale in different depths.

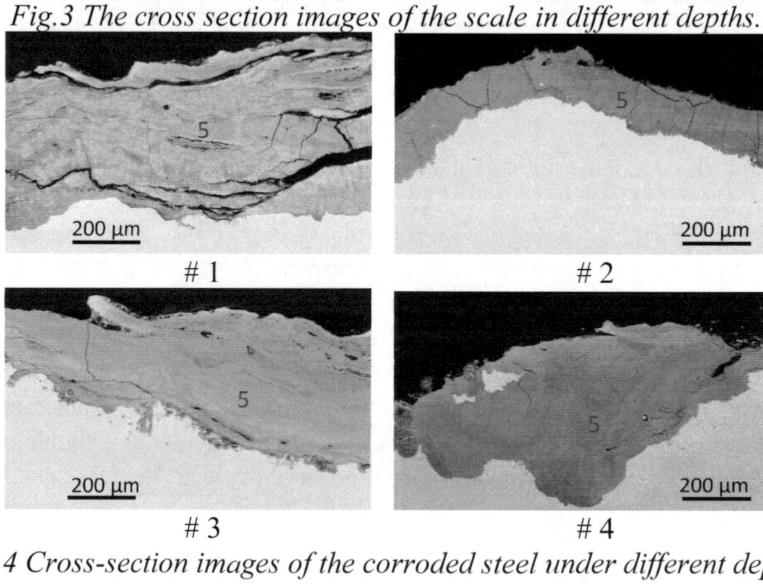

Fig.4 Cross-section images of the corroded steel under different depths.

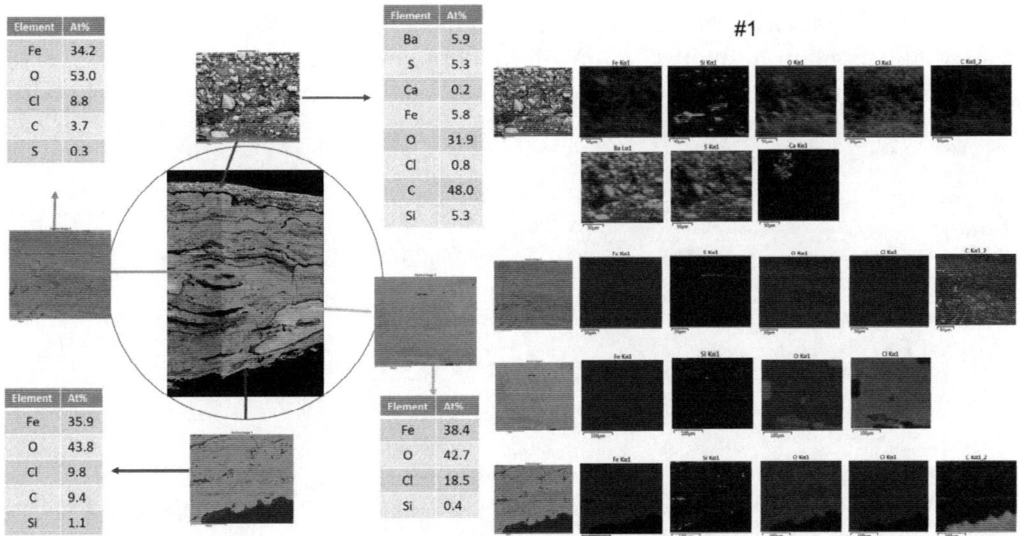

Fig.5 EDS images of scale # 1

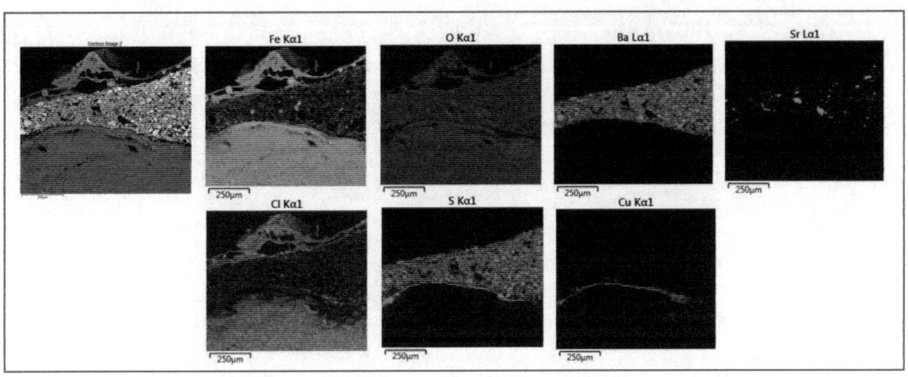

Fig.6 EDS images of scale # 4

The top part of the scale showed a layer of copper sulfide between the granular layer and bulk layer (Fig.6). It was a thin black layer, which was also found for samples from different depths. Table 2 showed the elements and possible phases of different parts based on the above-mentioned analysis.

Table 2 Contained element and possible phase in different layers of scale

	Elements contained	**Possible phase**
0	O, Fe, Cl, C	FeOOH
1	Ba, O, S, Ca, Si, C, Cl, Fe, Cu	$BaSO_4$, $CaSO_4$, SiO_2, $SrSO_4$, FeOOH
2	Ba, O, S, Cl, Fe, Cu	FeOOH, $BaSO_4$, Cu_2S
3	O, C, Fe, Cl, Si, Ca, Cu	FeOOH, SiO_2, $CaCO_3$
4	O, C, Fe, Cl, Si, Cu	Fe_xO_y, FeOOH, SiO_2
5	O, C, Fe, Cl, Si, Cu	Fe_xO_y,FeOOH, SiO_2

Chloride ions were easily transporting through the layer of corrosion scales and they were incorporated into the layer of scale and form iron hydroxide, akaganeite, β-FeO(OH) [18-19]. Figure 6 shows XRD diagram of the crystalline phases identified in the corrosion scale. The results showed that the main component of the scale was akaganeite and the chemical formula of akaganeite that explicitly indicates the presence of chloride ions in the structure is $Fe_8O_8(OH)_8Cl_{1.35}$. These ions had caused the formation of pitting and accelerated corrosion of steel and form an uneven surface. The second phase was the barite and strontium sulfate, which may also contain a small amount of calcium sulfate based on the EDS results. The scale may also contain minor amounts other forms of iron oxides or hydroxides, which had not be detected in the XRD spectrum.

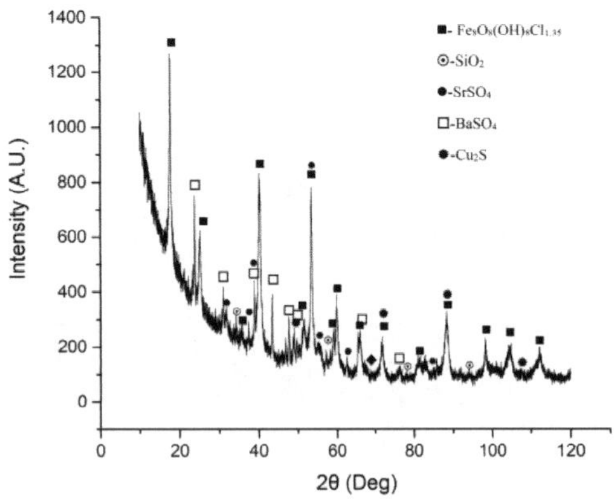

1#

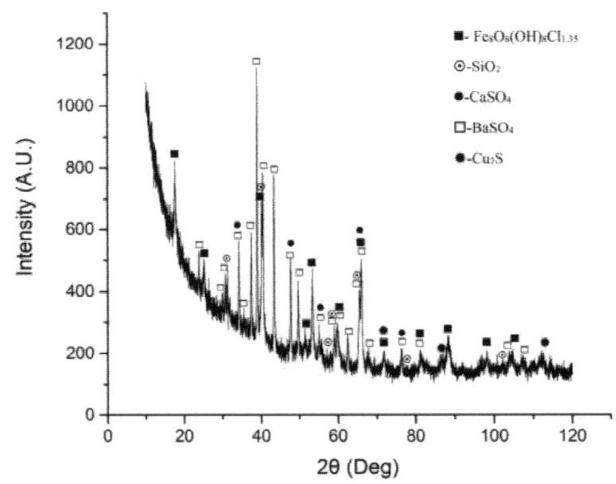

2#

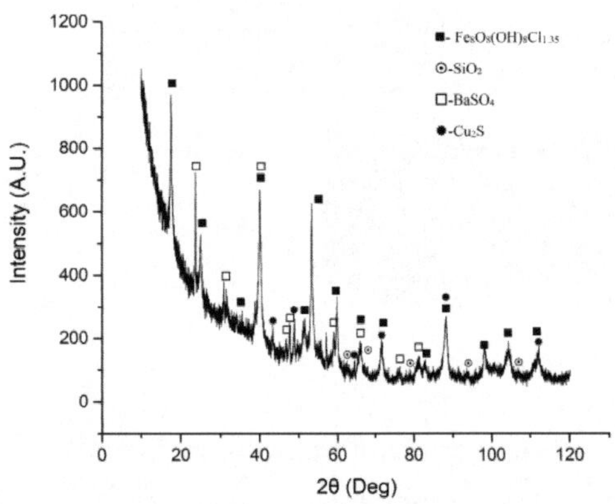

3#

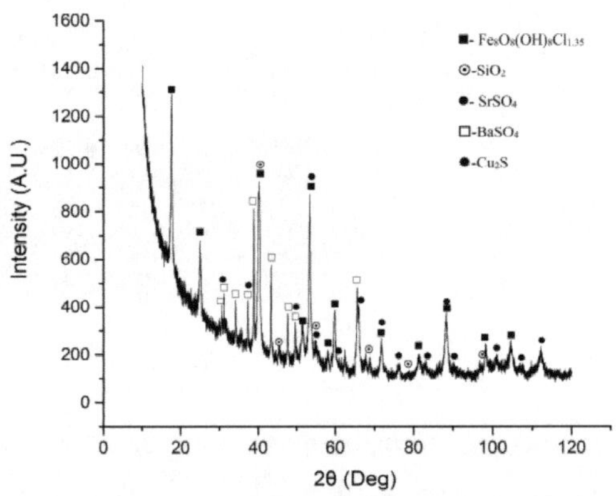

4#

Fig.7 XRD results of corrosion scales under different depths.

Formation-Growth Model

The iron scale formation and growth was influenced by several factors that include water quality parameters, flow patterns, temperature, microbiological activity and application of inhibitors. Therefore, the characteristics of iron scale in wellbore were unique for each site. However, scale formation and growth can be described by a reasonable hypothesis due to the scale come from different sources also have similarities. The composition and structure of iron scale in oil produced wellbore had been schematized and was presented as figure 8. The layered structure of the corrosion scale was obvious, with a dense layer of iron oxy-hydroxide, a porous layer of iron hydroxide and chloride rich products, a bulk layer of akagneite, a thin layer of Cu_2S and a granular layer of mixed sulfate, which had contained barite, strontium sulfate, and calcium sulfate.

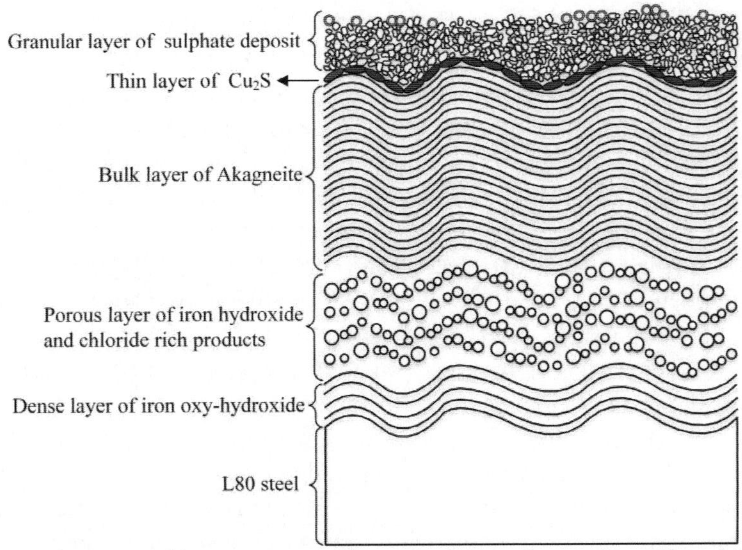

Fig.8 Schematic of iron scale structure and composition formed in oil produced wellbore.

A conceptual model had been developed to describe the formation and growth of iron scales, based on the history of the well. The process of iron scale formation and growth had be schematized, as shown in Figure 9. During the initial stage of corrosion, oxygen-rich water had reached the clean pipe surfaces and anodic corrosion reaction (dissolution) of steel ($Fe \rightarrow Fe^{2+} + 2e^-$) had taken place. Reactions of Fe^{2+} ions with water (hydrolysis) formed local acidic environment and produced the H+ ions. In this way, the anodic reactions (metal dissolution) and cathodic oxygen reduction occurred, as shown in Fig.9 (a). At this point, a variety of iron oxides or hydroxides may be formed, which had the low degree of crystallinity, indicating that the amorphous phases were present as iron oxy-hydroxide, δ-FeO(OH), which cannot be identified using XRD method. As the reaction continues, in order to establish an electrical balance with the positive ions produced on the surface of steel, the negative Cl^- ions had diffused from the surrounding environment into the layer of corrosion products [20]. When the concentration of Cl^- ions reached a critical value, it provided the necessary conditions for the formation of β-FeOOH. β-FeOOH had a strong reducing activity, which promoted the corrosion process. High concentration of Cl^- ions in the corrosion products stimulated the formation of akaganeite. After that, with the prolonged corrosion time, the corrosion product composition and relative content no longer significantly changed. As the scale thickness increased, it became increasingly difficult for the oxidant to diffuse to the metal-water interface. Different structural features of the corrosion scales started forming during this intermediate stage, as shown in Fig.9 (b).

The formation mechanism of the granular layer was not related the corrosion reaction. It was caused by the deposit of the sulfate, as shown in Fig.9 (c). Scale deposition was one of the most serious oil field problems when the sea water and the produced water were involved. For this well, seawater injection was used to stimulate production, seawater with the high concentration of SO_4^{2-} ion and formation water with high concentrations of Ca^{2+}, Ba^{2+} and Sr^{2+} ions. Mixing of these waters had caused deposition of $CaSO_4$, $BaSO_4$, and $SrSO_4$. There were other reasons why the scale formed and the amount and location of which were influenced by several factors. However, super-saturation was the most important reason behind mineral deposition. A supersaturated was the primary cause of scale formation and occurred when a solution contains dissolved materials that were at higher concentrations than their equilibrium concentration. In this model, the crystallization forms of barium sulfate and others were not the same as those commonly seen. Usually, the barite deposited was highly crystalline and large, and had an average size of about 10 μm. However, In granular layer, barite was poorly-crystalline, and the size of these particles was on the order of micrometer or sub-micrometer. Studies had shown that turbulence affects the crystalline state of barite. Regardless of whether inhibitor was present or not, turbulence or shear rate had a significant impact on the morphology of the barite deposit [21-22]. Therefore, it indicated that the fluid inside the well was turbulent when the deposition occurred.

After deposits were formed, it was difficult to stop under deposit corrosion, as shown in Fig.9 (d), because the deposits made it difficult to get corrosion inhibitors to the metal surface suffering the high corrosion rates. The water still presented inside the pores or cracks of the scales serves as the electrolyte for transport of ions for maintaining electroneutrality within the scale. Therefore, porous and conductive scales were least protective towards further corrosion of pipes. Moreover, when oxidants were not present in water, corrosion of the metal pipe may be sustained by reduction of corrosion scale components. The structure of the scale determined the ease with which iron can diffuse out of the scale in response to water quality conditions. Important reactions of the corrosion scales that resulted in the iron release.

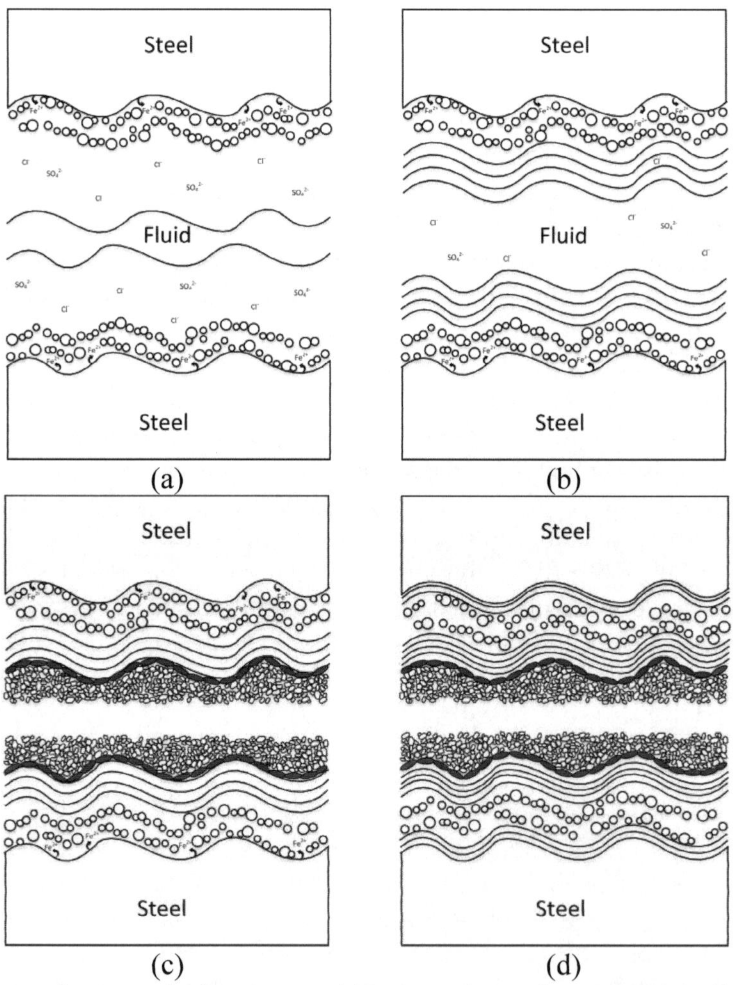

Fig.9 Schematic of iron scale formation: (a) porous layer formed; (b) bulk layer formed; (c) granular layer formed; (d) dense layer formed.

There was also a layer of copper sulfide between the granular layer and the bulk layer. The possible formation reasons were discussed below. Oil produced wellbore were very corrosive environments with a high concentration of chloride ion (Cl^-) and hydrogen sulphide (H_2S), the amount of Cl^- and H_2S in this wellbore was shown in Figure 9. Generally, Cr, Ni, Cu elements had been added to steel to enhance steel's corrosion resistance in H_2S and Cl^- environments [23-24]. The copper content in the sample of L80 steel was 0.35% wt. Copper was found in all layers except the granular layer, so that the copper in this layer came from the steel and the sulphur came from the dissolved H_2S. The thermodynamic calculation indicated that the copper sulphide was able to exist stably compared to the iron sulphide and the nickel sulphide [25]. Furthermore, Ueda et al. [26] have suggested that dissolved H_2S cannot pass through a sulfide layer which is formed in the H_2S and Cl^- environments, which means that the copper sulfide is able to reduce the activity of sulfur species on the inside of the sulfide. Therefore, it was difficult for the iron sulfide to form on the inside of copper sulfide due to the low activity of sulfur species, the iron oxide or iron hydroxide which does not contain the iron sulfide would form under copper sulfide.

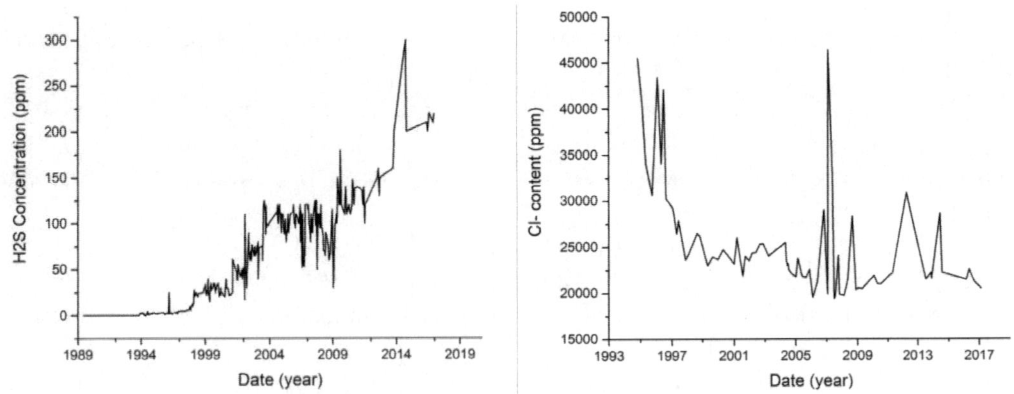

Fig.10 H_2S concentration and Cl^- content in oil produced wellbore.

Conclusions

The scale structure and composition played important roles in the migration of chloride ions and the reactions of corrosion scales that lead to iron release. Iron corrosion scales were layered deposits, with a dense layer of iron oxy-hydroxide, a porous layer of iron hydroxide and chloride rich products, a bulk layer of akagneite, a thin layer of Cu_2S and a granular layer of barite, strontium sulfate and calcium sulfate. Scale growth was a result of continued corrosion of the pipe, and continued in the presence of oxygen, with oxygen reduction taking place on the scale surface or inside the scales. The negative Cl^- ions diffused from the surrounding environment into the layer of corrosion products in order to establish an electrical balance with the positive ions produced on the surface of the steel. The main component of the scale was akaganeite and the chemical formula was $Fe_8O_8(OH)_8Cl_{1.35}$. A conceptual model had been developed to describe the formation and growth of iron scales, based on the history of the well. The crystallization of barite was affected by fluid, and when turbulent flow occurred, barite was poorly-crystalline, and the size of these particles was on the order of micrometer or sub-micrometer. The copper sulphide was able to exist stably compared to the iron sulphide, and reduced the activity of sulfur species on the inside of the sulfide. Chlorine ions played a key role in iron scale formation. After deposits were formed, it was difficult to stop under deposit corrosion, and the corrosion was probably worse because the thick scale provided a good environment for pitting corrosion.

Acknowledgments

The authors would like to thank The Danish Hydrocarbon Research and Technology Centre (DHRTC) for funding and Lars Pedersen, Steffen S. Munch, Flemming Jensen for their assistance and discussion in preparing samples and component analysis.

References

[1] Sanjiv Kumar, Tarun Kumar Naiya, Tarkeshwar Kumar, Journal of Petroleum Science and Engineering, 169(2018) 428-444.
[2] Mohammad Mohsen Vazirian, Thibaut V.J. Charpentier, MÃ´nica de Oliveira Penna, Anne Neville, Journal of Petroleum Science and Engineering,137(2016)22-32.
[3] Sassan Hajirezaie, Xingru Wu, Catherine A. Peters, Journal of Natural Gas Science and Engineering, 39(2017)188-202.
[4] Oscar Vazquez, Ilya Fursov, Eric Mackay,Journal of Petroleum Science and Engineering, 147(2016)302-307.
[5] Guolin Jing, Shan Tang, Xiaoxiao Li, Huaiyuan Wang, Arabian Journal of Chemistry, 10(2017) S1235-S1239.
[6] Arash Kamari, Farhad Gharagheizi, Alireza Bahadori, Amir H. Mohammadi, Fluid Phase Equilibria, 366(2014) 117-126.
[7] Nathan A.S. Webster, Bee K. Gan, Iztok Livk, Fuel,137(2014) 211-215.
[8] M.S. Al-Masri, A. Aba, Applied Radiation and Isotopes, 63(2005)457-463,
[9] Oscar Vazquez, Pablo Herrero, Eric Mackay, Myles Jordan, Journal of Petroleum Science and Engineering,138(2016)1-10.
[10]Azizollah Khormali, Dmitry G. Petrakov, Mohammad Javad Afshari Moein, Journal of Petroleum Science and Engineering,147(2016)843-850.
[11] Fan Yang, Baoyou Shi, Junnong Gu, Dongsheng Wang, Min Yang, Water Research, 46(2012) 5423-5433.
[12] Tammie L. Gerke, Kirk G. Scheckel, J. Barry Maynard, Science of the Total Environment, 408(2010)5845-5853.
[13] Peng Jin, Srdjan Nesic, H. Alan Wolf, Corrosion 2014, San Antonio, Texas, USA, NACE(2014)4075.
[14] Robert E. Melchers, Corrosion Science, 81(2014)110-116.
[15] S.M. Hoseinieh, A.M. Homborg, T. Shahrabi, J.M.C. Mol, B. Ramezanzadeh, Electrochimica Acta,217(2016) 226-241.
[16] Ali Haghtalab, Mohamad Javad Kamali, Abbas Shahrabadi, Fluid Phase Equilibria, 373 (2014)43-54.
[17] Yunjiao Fu, Wolfgang van Berk, Hans-Martin Schulz, Journal of Petroleum Science and Engineering, 112(2013)273-283.
[18] Selwyn, L.Overview of archaeological iron, the corrosion problem, key factors affecting treatment, and gaps in current knowledge, in: Proceedings of Metal 2004, National Museum of Australia, Canberra, Australia. (2004) 4–8
[19] Reguer, S., Mirambet, F., Dooryhee, E., Hodeau, J.,Dillmann, P. and Lagarde, P, Corrosion Science, 51(2009) 2795-2802.
[20] Tamura, H, Corrosion Science, 50 (2008)1872-1883.
[21] Fei Yan, Zhaoyi Dai, Gedeng Ruan, Hamad Alsaiari, Narayan Bhandari, Fangfu Zhang, Ya Liu, Zhang Zhang, Amy Kan, Mason Tomson, Journal of Petroleum Science and Engineering,149(2017)183-192.
[22] Fei Yan, Narayan Bhandari, Fangfu Zhang, Gedeng Ruan, Zhaoyi Dai, Ya Liu, Zhang Zhang, Scale Formation and Control Under Turbulent Conditions, SPE(2016)179863.

[23] K.B. Persson, S.J. Kaukas, U.H. Kivisäkk, Corrosion Performance of Alloy 29 in Simulated Sour Environments, Corrosion(2010), NACE: Houston, TX, Paper No.10317.
[24] M. Sagara, Y. Otome, H. Amaya, S. Kimura, M. Igarashi, M. Ueda, Development of High-strength Ni Alloy OCTG Material for Sour Environment, Corrosion(2011), NACE: Houston, TX, Paper No. 11109.
[25]Akiko Tomio, Masayuki Sagara, Takashi Doi, Hisashi Amaya, Nobuo Otsuka, Takeo Kudo, Role of alloyed copper on corrosion resistance of austenitic stainless steel in $H_2S–Cl^-$ environment, Corrosion Science, 81(2014)144-151.
[26] M. Ueda, T. Kudo, Effect of Alloyed Element on Corrosion Resistance of Ni Base Alloy in Sulfur Containing Sour Environment, Corrosion (1994), NACE: Houston, TX, Paper No. 69.

Morpho-chemical investigations and thermodynamic study of Ti, Nb rich nodules grown on AISI 441 oxidized at 800°C in wet atmosphere

Valerie PARRY[1], Wichitra WONGPROMRAT[2], Laurence LATU-ROMAIN[1], Céline PASCAL[1], Walairat CHANDRA-AMBHORN[2], Somrerk CHANDRA-AMBHORN[3], Yves WOUTERS[1]

[1] *University Grenoble Alpes, France*

[2] *King Mongkut's Institute of Technology Ladkrabang, Thailand*

[3] *King Mongkut's University of Technology North Bangkok, Thailand*

Solid oxide fuel cells produce clean and renewable energy. Their operating temperatures reduced to the intermediate range of 600 – 900 °C enable the use of metallic materials for interconnects. The poor creep resistance of ferritic stainless steels at SOFCs operating temperatures has urged the development of commercial alloys such as AISI 441. This steel contains small amounts of Ti (0.13 wt %) and Nb (0.55 wt %) which allow to stabilize its ferritic structure and to enhance the mechanical properties at high temperature.

As-received AISI 441 plates were oxidized from 4 to 24 h at 800 °C in wet atmosphere (5% H_2O in O_2). Micrometric oxide nodules containing Ti and Nb were observed to form on the passive scale. The development in service of the passive $Cr_2O_3/(Mn,Cr)_3O_4$ duplex oxide scale on ferritic stainless steels is well understood, as well as the possible development of iron containing oxide nodules leading to breakaway oxidation. However, the growth mechanism of Ti and Nb rich nodules and their possible mutation to catastrophic ones remain unclear. Since the composition and the morphology of the nodules are heterogeneous, characterizations were performed with complementary techniques: scanning transmission electron microscopy and energy dispersive X-ray (STEM-EDX) and dual focused ion beam scanning electron tomography (3D FIB-SEM). Eventually, the chemical compositions of the nodules and of the oxidation affected zone in the nodules vicinity are discussed in relation with thermodynamic calculations investigating the stability of the different oxides.

Experimental results reveal that nodules and their oxidation affected zones present a complex microstructure related to the presence of Si and Nb and their competition for interfacial oxidation. The metal/nodule interface is characterized by cavities as well as evidence of steel repassivation. Si enriched Laves phase particles are observed along this interface. Preferential oxidation of Si rather than Nb happens and results in the formation of a rather continuous SiO_2 subscale along the nodule edge, and the Laves phase particles.The nodules present a two parts structure. The inner part is embedded in the bulk alloy and is characterized by outgrowths composed of small oxide grains containing either Nb and Ti or Cr oxide enriched in Mn or Ti. The upper part is connected to the regular oxide scale. The top of nodules is porous and mainly composed of large Ti oxide grains together with Mn enriched Cr oxide grains. It does not contain Fe excepted as Fe oxide filaments and as Fe enriched Cr, Mn oxide areas localized along the pores. Comparing EDX results which show very few oxidised Fe in the nodules and the thermodynamic calculations where Fe is largely present in corundum and spinel, it is concluded that the nodules described in this paper present a kind of passivity. They cannot be compared with the Fe-rich ones emerging during breakaway oxidation.

Hot corrosion of HP/HVOF sprayed MCrAlY coatings in molten salt environment

Kate¿ina KOPELENTOVÁ[1], Zdeněk ČESÁNEK[1], Jan SCHUBERT[1]

[1] *Výzkumný a zkušební ústav Plzeň s.r.o., Czech Republic*

The present study is focused on investigation of corrosion behaviour of the MCrAlY coatings in molten salt environment. Hot corrosion test were conducted on the HP/HVOF sprayed NiCoCrAlYHfSi, NiCoCrAlY, NiCoCrAlTaReY and CoCrAlYTaCSi coatings deposited on chromium-molybdenum steel X10CrMoVNb91. The specimens were exposed to aggressive environment of molten salt 40% Na_2SO_4 60% V_2O_5 at 750 °C under cyclic conditions. The thermogravimetric technique was used for evaluation of corrosion kinetics. The corrosion products were analysed using SEM, EDS and XRD methods. Microhardness was measured for as-sprayed as well as exposited specimens. The NiCoCrAlTaReY a NiCoCrAlY coatings have high corrosion resistance and the NiCoCrAlYHfSi coating was shown the lowest hot corrosion resistance in the given molten salt environment.

High temperature hot corrosion behavior of selected thermally sprayed coatings in an aggressive environment at 690 °C

Zdeněk ČESÁNEK[1], Kateřina KOPELENTOVÁ[1], Jan SCHUBERT[1]

[1] *Výzkumný a zkušební ústav Plzeň s.r.o., Czech Republic*

This study is focused on hot corrosion evaluation of selected thermal sprayed coatings. Cr_3C_2-25%NiCr, Stellite 6, NiCrBSi and Hastelloy C-276 was sprayed to the substrate material 1.4923. This material is used as an efficient construction material for components of power equipment. All tested coatings were sprayed using HP / HVOF (High Pressure / High Velocity Oxygen Fuel) technology.

Hot corrosion studies were performed in a molten salt environment typical for coal-fired boilers. All coatings was exposited Na_2SO_4 82 % $Fe_2(SO_4)_3$ environment at 690 °C under cyclic condition. Each cycle consisted 1h heating in silicon carbide tube furnace followed by 20 min of cooling. The weight change measurement was performed after each cycle to establish the corrosion kinetics. Scanning electron microscopy (SEM), elemental analysis (EDS) and optical microscopy (OM) technique were used to analyse the corrosion mechanism.

The effect of anchor group and backbone chain on performance of organic compounds as corrosion inhibitors for aluminium investigated using integrative experimental-modeling approach

Ingrid MILOSEV[1], Dolores ZIMERL[1], Sandrine ZANNA[2], Antoine SEYEUX[2], Jernej ISKRA[1], Stojan STAVBER[1], Matic POBERZNIK[1], Aniko NEMES[3], Dominique COSTA[2], Jozsef RABAI[3], Anton KOKALJ[1], Philippe MARCUS[2]

[1] *Jozef Stefan Institute, Slovenia*

[2] *CNRS Chimie ParisTech, France*

[3] *Eotvos Lorand University, Hungary*

The use of corrosion inhibitors is one of the oldest method of corrosion protection of metals and alloys. Numerous different compounds are being used for this purpose; they may be added to medium within closed systems like pipelines and cooling systems, or added to coating, primer or topcoat aiming to contribute to additional corrosion protection. Current state-of-the-art in this field is represented by high throughput experimental empirical testing of large sets of organic compounds. In contrast to such trial and error approach, a rational design of new corrosion inhibitors would provide a major breakthrough in the field of corrosion protection. The scientific framework should therefore postulate the strategy which will contribute to the increased understanding of corrosion inhibition. One of the possibility it to use the ICME (Integrated Computational Materials Engineering) principles. This work therefore addresses the basic relationships between aluminium surface and corrosion inhibitors, i.e. corrosion inhibitors were designed in such a way to allow the investigation of the effect of anchor group and the effect of backbone chain on their inhibitive performance on aluminium in chloride solution. Methodological approach was versatile: (i) synthesis or purchase of organic inhibitors, (ii) preparation of uncoated and inhibitor-coated aluminium samples, (iii) electrochemical measurements, (iv) water contact angle measurements (v) long-term immersion tests, (vi) surface analytical study by SEM, ToF-SIMS or XPS, and (vii) Density Functional Theory (DFT) modeling. The surface of aluminium substrates was prepared by alkaline etching to allow easy self-assembling during immersion in ethanol solution of selected corrosion inhibitors. The inhibitors differed in the (i) type of backbone chain (alkyl and perfluoro), (ii) length of backbone chain, and (iii) type of anchor group (carboxylic, azide, imidazole, thiol, dioctyl, thiocyanate, phosphonic and benzoic).

Based on the obtained experimental (electrochemical parameters and composition of the surface layers) and computational (adsorption reaction energies decomposed into inhibitor-surface and lateral inhibitor-inhibitor interactions, adsorption induced changes of electronic structure, etc.) parameters we are aiming to build a predictive model for screening new corrosion inhibitors with targeted properties. The modeling and experiments are therefore combined by an integrative approach.

This work is a part of M-ERA.NET project entitled "Design of corrosion resistant coatings targeted for versatile applications" (acronym COR_ID). The financial support of the project by MESS (Ministry of Education, Science and Sport of Republic of Slovenia), ANR (The French National Research Agency) and NKFIH OTKA (National Research Development and Innovation Office, Hungary) is acknowledged. Région Ile-de-France is acknowledged for partial support of the ToF-SIMS equipment.

Gas phase conductivity and its effect to Cr2O3 volatility in Ar and Ar-5%H2

Syamsul Kamal ARIFIN[1], Ahmad Nukhaie BERAHIM[1], Mohd Hanafi ANI[1]

[1] *International Islamic University Malaysia, Malaysia*

Fe-Cr alloy is commonly being used as boiler tube's material. It is subjected to prolonged exposure to water vapor oxidation. The ability to withstand high temperature corrosion can normally be attributed to the formation of a dense and slow growing Cr-rich-oxide scale known as chromia, Cr_2O_3 scale. This scale protects Fe-Cr from further high temperature corrosion. However, oxidation may limit the alloy's service lifetime due to declining of its protective capability. One of the factors to this issue is hydrogen induced cracking which paradoxically, water vapor dissociates to dissolved hydrogen. Thus, these led to the need of studying the effect of dissolved hydrogen on high temperature resistance of chromia scale. This work is giving special attention on the effect of gas phase of Ar and Ar-5%H_2 conductivity to Cr_2O_3 volatility. A dense sintered Cr_2O_3 was being exposed to dry (Ar) and wet (Ar-5%H_2) environment for 172.8 ks (48 h) at 1073 K. FT-IR result confirms that dissolved hydrogen affect the chemical bonding of Cr_2O_3 and thus later promotes volatilization of protective Cr_2O_3 scale. I-V measurement in wet and dry gases gives varying degree of conductivity, most probably due to electron exchange. The gas conductivity behaves according to Ohmic law, steeper in wet condition (with present of hydrogen). The result provides an indication to corrosiveness of Cr_2O_3 scale. It was found that resistance ratio, value is around 1.4. It is concluded that the presence of hydrogen in gas phase contributes to 40% higher corrosivity.

Assessment of Glycine Propyl Ester Lauryl Hydrochloride as Sustainable Mild Steel Corrosion Inhibitor in 1M HCl

Saman ZEHRA[1], Mohammad MOBIN[1]

[1] *Aligarh Muslim University, India*

Corrosion is a severe challenge which restricts the use of Mild Steel (MS) in various applications where it comes in contact with aggressive chemicals and conditions. In this regard, different inhibitors were chosen, but environment unfriendliness of such inhibitors is a primary cause of concern. The present work describes the corrosion inhibiting properties of the synthesized amino acid derivative i.e., glycine propyl ester hydrochloride on the corrosion of MS in 1M HCl by employing gravimetric, electrochemical measurement techniques. EIS studies indicated the formation of adsorbed shielding film on the steel/electrolytic solution interface by the inhibitor molecules whose existence was affirmed by SEM/EDX. FT-IR spectroscopic analyses affirmed the presence of chemical interactions between the inhibitors and MS surface. The evaluated compound works as an effective inhibitor for acid corrosion at substantially lower concentration and its adsorption on the MS surface was found to follow the Langmuir adsorption isotherm. Calculated thermodynamic parameters for adsorption unveiled a strong interaction amongst the inhibitor-MS surface. The electrochemical results revealed that the inhibitor act as mixed-type. Theoretical calculations using DFT were accomplished to ascertain the mechanism of corrosion inhibition.

Impact of thermal cycles on the stability of grown oxide layers on 9 – 12 % Cr steels under water steam

Maria MOSQUERA FEIJOO[1], Axel KRANZMANN[1], Jürgen OLBRICHT[1]

[1] *Federal Institute for Materials Research and Testing, Germany*

Flexible operating conditions in power plants include cyclic thermal/mechanical loading. As a result the oxide scale of construction materials might fail by formation of cracks and/or spallation. The oxide spalls can cause steam tube blocking and overheating, which might result in drastically reduced component service life.

The aim of this work is to comprehensively analyze and predict the performance of ferritic-martensitic heat resistant steels (9 – 12 % Cr) under thermo-cyclic conditions. Different sample geometries (rectangular coupons, full rings and U-shape segments) were isothermally pre-oxidized at 620 °C, 30 bar for 1500 hours in pure steam to generate a defined oxide layer of certain thickness. After pre-oxidation, the samples were subjected to thermal cycling between 620 – 320 °C for up to 258 cycles in pure steam at 30 bar.

After the experiments, oxidation kinetics (in terms of mass change, scale thickness and material loss) and the impact of the geometry are investigated. The integrity of the oxide layers and their adherence to the substrate are evaluated.

Reactivity of 9-12Cr steel in CO2-O2 environment: comprehensive description

FABIEN ROUILLARD[1]

[1] *CEA, France*

Supercritical CO_2 (SCO_2) Brayton cycle is gaining more and more interest for nuclear and fossil power plants thanks to its possibility of improved efficiency combined with smaller turbo-machinery size compared to a traditional Rankine steam system. Below 600°C, 9-12Cr steels would be good candidates thanks to their good physical properties. Unfortunately, they usually suffer from poorer corrosion behaviour than usual austenitic steels [1]. Recent studies have suggested that the corrosion behaviour of 9-12Cr steels in CO_2 environment can be in fact strongly influenced by the partial pressure of O_2 impurity in CO_2 and that protective chromium rich may form in certain conditions [2]. In order to better understand the influence of O_2 impurity in CO_2 on the corrosion behaviour of 9-12Cr steels, the reactivity of 9Cr steel in CO_2/O_2 mixture has been studied in a specific device dedicated to the understanding of gas-solid reaction. The purpose of this device is to identify the gas-solid reaction mechanism by using mainly labelled molecules, mass spectrometer and pressure gauges. From this study, it is demonstrated that the oxygen molecules contained in CO_2 as impurities react preferentially in the first instants of oxidation and that their partial pressure determines the chromium enrichment of the formed oxide layer.

References

[1] F. Rouillard, F. Charton, G. Moine, Corrosion behavior of different metallic materials in supercritical carbon dioxide at 550°C and 250 bars, CORROSION vol 67 n°9 (2011)

[2] Bouhieda et al, Materials at High Temperatures 29 (2) (2012)

Aqueous Corrosion through Galvanic Coupling in a Two-Phase Material: The Role of Interface Crystallography

Irshad KHAN[1]

[1] *IIT BOMBAY, India*

Aqueous corrosion, through galvanic coupling, in a two-phase eutectoid steel was shown to be restricted to the ferrite side of the ferrite-cementite interface. The interface boundary density, or interface length per unit area, and the interface crystallography, orientation relationship between ferrite and cementite, determined the corrosion current and average depth of attack. By suitable microstructural tailoring, corrosion rates were thus significantly altered

Application of Vapor Phase Corrosion Inhibitors for Contaminated Environments

Behzad BAVARIAN[1], Lisa REINER[1], Boris MIKSIC[2]

[1] CSUN-MSEM, United States of America

[2] Cortec Corp., United States of America

Availability and accessibility to clean water for industrial applications such as hydrotesting and hydrostatic testing of pipelines, vessels, and plumbing distribution systems has been a challenging task especially when the job sites are in remote locations. These circumstances have dictated the use of contaminated water containing for example, large amounts of chloride. Hydrotesting or hydrostatic testing of metallic equipment is an important method for quality control of parts, after production and during use. When contaminated water is used in these tests, corrosion of the metallic components of the system, during and after testing become a major concern, and application of inhibiting system is required. Vapor-phase corrosion inhibitors (VCI), when added to water in small concentrations are able to prevent corrosion during testing, during wet storage and after testing. These inhibitors are to prevent corrosion of metallic components in three phases: within the solution, at the water line, and above the water line. VCI-based inhibitors prevent corrosion in the areas that are not accessible by other corrosion inhibitors.

Several commercially available corrosion inhibitors were evaluated in this investigation to optimize the best combination of inhibitors to minimize the corrosion to these multi-metallic components system. Laboratory studies are presented in this paper to show the effectiveness of VCI products in both fresh and salt water applications.

Application of an inhibitor in any water system requires compliance with the Clean Water Act, the National Pollutant Discharge Elimination System (NPDES). The corrosion inhibitors investigated were water-based, biodegradable, environmentally friendly, and can be effective replacements for toxic nitrite, chromate or hydrazine base inhibitors.

Key Words:

Hydrotesting, vapor-phase corrosion Inhibitors, environmentally friendly inhibitor, biodegradable inhibitor

Microscopic investigations of a Cr2O3 scale formed on a substrate with implanted yttrium ions

<u>Aleksander GIL</u>[1], Oleksandr KRYSHTAL[2], Tomasz BRYLEWSKI[1], Aleksandra CZYRSKA-FILEMONOWICZ[2]

[1] *AGH University of Science and Technology, Faculty of Materials Science and Ceramics, Poland*

[2] *AGH University of Science and Technology, International Centre of Electron Microscopy for Materials Science and Faculty of Metals Engineering and Industrial Computer Science, Poland*

Atomic-resolution (S)TEM methods were used to examine the structure of a Cr_2O_3 scale formed on chrome and the Co-40Cr alloy, both of which had undergone yttrium ion implantation at a dose of 2´10^{16} at./cm^2 and a beam energy of 70 keV. Oxidation was performed for 100h in air at 800 °C.

Cross-section lamellae for (S)TEM investigations were prepared using focused ion beam (FIB) NEON CrossBeam 40EsB SEM from ZEISS. A layer of platinum was deposited on the surface of the scale, and lamellae with the dimensions of 7´10 μm^2 and a thickness of around 80 nm were then cut from the scale in the perpendicular direction. Microscopic examinations were conducted using the S/TEM Titan Cubed G2 60-300 transmission electron microscope equipped with the ChemiSTEM (FEI) system. Phase identification was performed by means of selected area electron diffraction (SAED). STEM images were observed by means of high-angle annular dark-field detectors. The chemical composition was analyzed using energy dispersive X-ray spectrometry EDS).

The microscopic observations of the scale cross-section revealed that it is composed of micrometre-sized Cr_2O_3 crystallites with flat walls. The STEM-EDS and SAED analyses showed that in the case of both substrates yttrium was located in the interior of the Cr_2O_3 scale, in the form of a layer composed of the $YCrO_3$ perovskite. On the other hand, the presence of yttrium was not observed either in the interior of Cr_2O_3 grains or at the scale/substrate interface.

If the $YCrO_3$ is to be treated as a marker, then its location would suggest that the Cr_2O_3 scale formed on a substrate with implanted yttrium ions grows predominantly as a result of the predominant inward diffusion of oxygen across points defects in the anion sublattice, with a certain contribution of the outward diffusion of chromium.

Inhibitive properties of sodium gluconate and biocide (CTAB) on corrosion, biocorrosion and scaling controls of copper in cooling water systems

Mohamed EBN TOUHAMI[1], Elhouseine HOUSNI[2], Nadia DKHIRECHE[2]

[1] *Faculty of Science - University Ibn Tofail, Morocco*
[2] *Faculty of science, Morocco*

Inhibitive properties of sodium gluconate and biocide (CTAB) on corrosion, biocorrosion and scaling controls of copper in cooling water systems

Elhousni[1], N. Dkhireche[1], M. Ebn Touhami[1]*, M. Galai[1], R. Touir[1]

[1]Laboratory of Materials Engineering and Environment: Modeling and Application, Faculty of sciences, Ibn Tofail University, Kenitra, Morocco

*Corresponding author: E-mail:

Abstract

The use of water as thermal fluid in cooling circuits usually leads to three phenomena namely: Corrosion, scaling and biological fouling, which constitute the cause of a decrease in thermal and electrical efficiency of water circuits, a loss of metallic material and the proliferation of microorganisms in the water circuits.

In this study, the characterization of the studied water was effectuated and has revealed that it is corrosive, scaly and seat of development of the microorganisms. Then researches of new formulations have been started in order to propose effective solutions for protection of cooling circuits. First of all, our choice was made on sodium gluconate (GS), combined with cetyltrimethyl ammonium bromide biocide (CTAB) used to fight against the proliferation of microorganisms.

The investigation of obtained results by using electrochemical methods has demonstrated that GS is an efficient corrosion and scaling inhibitor with efficiency close to 95% at 10^{-3}M. Also the study has shown that CTAB biocide is a mixed type inhibitor with a good inhibitory efficiency of 86% at 15 ppm.

In addition, the optimization of GS and CTAB concentrations was carried out by an electrochemical study in order to find a formulation to prevent cooling circuits from the problems previously mentioned. The formulation found (10^{-3}M of SG + 15 ppm of CTAB) has a better performance and has an efficiency of the order of 89% which improves with the immersion time to reach 94% after 24 hours and it keeps its performance with higher temperatures.

The Scanning Electron Microscopy (SEM, Energy dispersive X-ray (EDX) and X-ray Diffraction (XRD) analysis have shown that this formulation is an efficient inhibitor for corrosion, scaling and biological fouling. These analyses confirm all results obtained by potentiodynamic polarization curves and electrochemical impedances spectroscopy.

Key words: cooling water system, copper, inhibition, corrosion, scaling, formulation, Microbiology, sodium gluconate, biocide.

Effect of water chemistry on scaling in heating systems

Georgeanna KAWALEY[1]

[1] *Mitsubishi Electric R&D Center Europe BV, United Kingdom*

The complexity of water chemistry and it's variation in distribution systems continues to pose challenges for manufacturers in the heating, ventilation and cooling (HVAC) systems industry. Constant monitoring is one reliable means of overcoming unforeseen issues relating to water composition in a heating system water circuit. Another is by understanding how certain components of water behave under specific conditions akin to system/ product operation.

Hard water which is water containing high calcium carbonate or magnesium carbonate content is the most well-known cause of scaling. In this paper, it is shown how other components of water can have a significant effect on scale formation and deposition in a heating system. By understanding such mechanisms, an advantage is gained in light of system manipulation to mitigate/ prevent scaling and in turn reduce the risks of system/ product degradation.

investigation of 4-Amino-2,3-dimethyl-1-phenyl-3-pyrazolin-5-one as corrosion inhibitors for carbon steel in 1M HCl solution

Ali HEIDARY MOGHADAM[1], **Mohamad AJAMIGHALEDARVISH**[1]

[1] *Dezful Branch, Islamic Azad University, Iran*

The influence of 4-Amino-2, 3-dimethyl-1-phenyl-3-pyrazolin-5-one was investigated as corrosion inhibitors of carbon steel in 1 M HCl solution using Tafel polarization and electrochemical Impedance spectroscopy techniques. The results showed that 4-Amino-2, 3-dimethyl-1-phenyl-3-pyrazolin-5-one inhibited the corrosion of carbon steel in 1M HCl. 4-Amino-2, 3-dimethyl-1-phenyl-3-pyrazolin-5-one suppressed both cathodic and anodic processes of steel corrosion in 1M HCl, 4-Amino-2, 3-dimethyl-1-phenyl-3-pyrazolin-5-one molecules via mixed adsorption mechanism. 4-Amino-2, 3-dimethyl-1-phenyl-3-pyrazolin-5-one was adsorbed on the carbon steel surface according to a Langmuir isotherm adsorption model.

Key words: inhibitors, carbon steel, Impedance spectroscopy

The synergistic inhibitive effect of polymer and gemini surfactant on the hydrochloric acid corrosion of mild steel

Ruby ASLAM[1], **Mohammad MOBIN**[1], **Jeenat ASLAM**[2]

[1] *Aligarh Muslim University, Aligarh, India*

[2] *2Department of Chemistry, College of Science, Taibah University, Yanbu, Al-Madina, Saudi Arabia*

The synergistic inhibitive effect of polymer and gemini surfactants on the hydrochloric acid corrosion of mild steel

Ruby Aslam[1], Mohammad Mobin[1], Jeenat Aslam[2]

[1]Corrosion Research Laboratory, Department of Applied Chemistry, Faculty of Engineering and Technology, Aligarh Muslim University, Aligarh, 202002

[2]Department of Chemistry, College of Science, Taibah University, Yanbu, Al-Madina

Abstract

The corrosion inhibition behavior of the polymer–surfactant in mixed solution of the synthesized green cationic i.e., ethane-1,2-diylbis(N,N-dimethyl-N-dodecylammoniumacetoxy)dichloride, gemini surfactant referred as (12-E2-12) and a polymer sodium carboxymethylcellulose (NaCMC) in HCl solution was studied by electrochemical technique, computational and MD simulation study. The studied system (12-E2-12/CMC) inhibited the corrosion of mild steel in 1M HCl solution and the extent of inhibition was dependent on concentration. 12-E2-12/CMC is mixed-type inhibitor, according to results of corrosion potential obtained from potentiodynamic polarization. The decreasing values of C_{dl} and increasing values of R_{ct} obtained during EIS measurements, suggests formation of protective layer over the mild steel surface. SEM micrographs displayed clearly different morphology in presence of studied inhibitory system confirmed the existence of an adsorbed protective film on the mild steel surface. Quantum chemical calculations and MD simulation were applied to elucidate the adsorption behaviour and inhibition mechanism of studied system.

Assessment of corrosion inhibitive behavior of 2-mecraptobenzothiazole and sodium phosphate dibasic on industrial steel pipelines

Ioannis KARTSONAKIS[1], Evaggelia KARAXI[2], Panagiota STAMATOGIANNI[2], Panagiotis KAVOURAS[2], Costas CHARITIDIS[2]

[1] *National Technical University, Greece*

[2] *RNanolab, School of Chemical Engineering, National Technical University of Athens, Greece*

Mild or low-carbon steel comprises the most widespread kind of steel, not only for structural purposes but also for various industry or more ordinary applications. Due to mild steel's increasing utilization, there has been great concern for the susceptibility of such steels in weather corrosive factors, like humidity, acidity or salinity of the atmosphere.

Evaluation of susceptibility to corrosion of steel in industrial hot/ cooling systems after their exposure in 5mM NaCl as well as in water supply system of the city of Athens was carried out in this work. The corrosion protection effectiveness of the complexes or oxides or salts that are formed on the metal alloy surface, as well as the morphological modifications that occur on the surface of the steel panels following exposure in the presence or absence of 2-mecraptobenzothiazole (MBT) and sodium phosphate dibasic (Na_2HPO_4), was evaluated by three families of techniques. Electrochemical techniques [electrochemical impedance spectroscopy (EIS), potentiodynamic polarization method and linear polarization resistance] were used for corrosion estimation of steel. Microscopy techniques [scanning electron microscopy (SEM) and atomic force microscopy] were used for collecting qualitative information on the surface damage and morphological changes that occur on the metal surface as a result of the corrosion process. Finally, spectroscopy methods [energy dispersive X-ray analysis spectroscopy and X-ray diffraction (XRD)] were conducted to evaluate the formation of potential complexes on the metal alloy surface following its interaction with the two considered corrosion inhibitors, in the corrosive environment.

The electrochemical characterizations demonstrated that MBT and Na_2HPO_4 can be considered as corrosion inhibitors of steel as they reduced the corresponding anodic and/or cathodic reactions. Moreover, the synergistic effect of the corrosion inhibition behavior of MBT and Na_2HPO_4 in molar ratio 1:1 at several concentrations was studied. The results revealed that the admixture performed effectively with inhibition efficiency above 90%. The presence of both aforementioned inhibitors into the corrosive environment exhibited the highest impedance modulus and R_p value as the immersion time elapsed. XRD measurements and SEM characterization depicted that the exposure of steel to corrosive environment in the presence or absence of inhibitors, resulted in the formation of various oxide, hydroxide and hydroxide-chloride compounds on the steel surface. Considering the electrochemical, spectroscopy and morphology characterizations, the corrosion protection mechanisms of steel can be attributed to the protective layers formed onto the metal surface due to the presence of the inhibitors, which prevent chlorides penetration.

The effect of reactive elements on the formation and hot corrosion performance of aluminide coating

Farhad SHAHRIARI NOGORANI[1], Farhad FATEMI[1]

[1] *Shiraz University of Technology, Iran*

Low concentrations of reactive elements (REs) such as Y and Ce can significantly improve the oxidation behavior of high temperature alloys and coatings. In this investigations pack cementation aluminide coatings were developed on a nickel alloy with additions of reactive element oxides to the coating powder mixture. The mixture was composed of 3 wt.% RE oxide (RE = Ce, Y, La and Zr), 2 wt. % ammonium chloride, 15 wt. % aluminum, and the reminder alumina. Aluminizing treatment was carried out at 750 °C for 2 h. The aluminized specimens were then diffusion annealed at 1050 °C for 4 h under argon atmosphere. Characterization of cross-sections of the specimens by means of SEM/EDS and XRD confirmed the formation of NiAl phase on all the specimens with the presence of low concentrations (0.1 to 1.5 wt. %) of REs in the coatings. However for all types of REs examined, the thickness of the RE-modified specimens was less than the un-modified coating formed in similar conditions. As a result, during subsequent hot corrosion testing, the expected positive effects of REs were obscured by the less aluminum content of the modified coatings. The hot corrosion test was carried out at 900 °C with about 1 mg/cm2 loading of Na2SO4-NaCl-V2O5 salt. After 100 h of hot corrosion testing the Zr-modified coating showed the best resistance in terms of the amount of Al loss induced formation of Ni3Al in the coating.

High inhibiting effect of selected carboxylates on magnesium corrosion

Sviatlana LAMAKA[1], Junjie YANG[1], Kiryl YASAKAU[2], Alina MALTSEVA[3], Di MEI[1], Carsten BLAWERT[1], Polina VOLOVITCH[3], Mikhail ZHELUDKEVICH[1]

[1] *Magnesium Innovation Centre, Helmholtz-Zentrum Geesthacht, Germany*

[2] *CICECO – Aveiro Institute of Materials, University of Aveiro, Portugal*

[3] *Chimie ParisTech, PSL Research University, CNRS, Institut de Recherche de Chimie Paris, France*

For decades, it has been known that minor impurities of transition metals, especially iron, nickel and copper, significantly accelerate corrosion rate of magnesium and its alloys [1]. However, the reason for this effect has not been fully understood until recently. In the past three years, it has been independently shown by several groups of researchers, that chemical dissolution followed by re-deposition of transition metals on magnesium increases the area of cathodic reaction, accelerating magnesium corrosion [2-5].

Following these findings, a new concept of magnesium corrosion inhibition has been proposed [6]. Binding dissolved Fe^{2+}/Fe^{3+} into the complex prevents iron re-deposition. Testing inhibiting efficiencies of over a hundred substances has shown that iron complexing agents are efficient inhibitors of magnesium corrosion [7-9].

Among the massive number of tested inhibitors, sodium salts of 2,5-pyridinedicarboxylic, 3-methylsalicylic and fumaric acids possess high inhibiting efficiencies (>90%) to three grades of pure Mg and six industrially relevant alloys [7-9]. Inhibiting mechanisms of 2,5-pyridinedicarboxylate, 3-methylsalicylate and fumarate have been investigated by a number of electrochemical, surface analysis and microscopy techniques including SKPFM and GDOES. The presentation will discuss the origin of high inhibiting effect and particular mechanism of each inhibitor.

[1] R.E. McNulty, J.D. Hanawalt. Some Corrosion Characteristics of High Purity Magnesium Alloys. Transactions of The Electrochemical Society, 81 (1942) 423-433.

[2] D. Höche, C. Blawert, S.V. Lamaka, N. Scharnagl, C. Mendis, M.L. Zheludkevich. The effect of iron re-deposition on the corrosion of impurity-containing magnesium. Physical chemistry chemical physics : PCCP, 18 (2015) 1279-1291.

[3] J. Li, W. Sun, B. Hurley, A.A. Luo, R.G. Buchheit. Cu redistribution study during the corrosion of AZ91 using a rotating ring-disk collection experiment. Corrosion Science, 112 (2016) 760-764.

[4] K.A. Yasakau, D. Höche, S.L. Lamaka, M.G.S. Ferreira, M.L. Zheludkevich. Kelvin Microprobe Analytics on Iron-Enriched Corroded Magnesium Surface. CORROSION, 73 (2017) 583-595.

[5] D. Mercier, J. Światowska, S. Zanna, A. Seyeux, P. Marcus. Role of Segregated Iron at Grain Boundaries on Mg Corrosion. Journal of The Electrochemical Society, 165 (2018) C42-C49.

[6] S.V. Lamaka, D. Höche, R.P. Petrauskas, C. Blawert, M.L. Zheludkevich. A new concept for corrosion inhibition of magnesium: Suppression of iron re-deposition. Electrochemistry Communications, 62 (2016) 5-8.

[7] S.V. Lamaka, B. Vaghefinazari, D. Mei, R.P. Petrauskas, D. Höche, M.L. Zheludkevich. Comprehensive screening of Mg corrosion inhibitors. Corrosion Science, 128 (2017) 224-240.

[8] S. Lamaka, D. Höche, C. Blawert, M. Zheludkevich. WO/2017/064185, 2017.

[9] S.V. Lamaka, D. Höche, M.L. Zheludkevich. EP16184196, 2016.

Corrosion acceleration by RE3+ ions

Marco OLIVEIRA[1], Silvar KALLIP[2], Alexandre BASTOS[1], Theodor HACK[3], Mikhail ZHELUDKEVICH[4], Mário FERREIRA[1]

[1] *CICECO, DEMaC, University of Aveiro, Portugal*
[2] *Institute of Chemistry, University of Tartu, Estonia*
[3] *Airbus Central R&T, Germany*
[4] *Magnesium Innovation Centre–MagIC, Helmholtz-Zentrum Geesthacht (HZG), Germany*

In the last years RE (rare earth) ions, particularly Ce^{3+}, have been extensively studied as corrosion inhibitor for a number of metals and alloys. The Ce^{3+} action is attributed to its precipitation with OH^- ions from the oxygen reduction reaction blocking the cathodic sites.

This communication shows a few examples of galvanic corrosion accelerated, not retarded, in the presence of RE^{3+} ions. The work was performed by monitoring the galvanic current with a ZRA (Zero Resistance Ammeter) and also by SVET (Scanning Electrode Vibrating Technique) and polarization curves.

It has been found that a link between the increase in current and the RE ion definitely exists, independent of the conjugate cation or the presence of oxygen. Nevertheless, the nature of the substrate is deeming to the nature of the response.

The EU FP7 project PIAPP-GA-2013-612415 (PROAIR) and FCT Exploratory project IF/00856/2013/CP1162/CT0019 are greatly acknowledged for financial support.

A look into the inhibition mechanisms of mixtures containing RE trivalent ions and organic compounds.

Marco OLIVEIRA[1], Silvar KALLIP[2], Alexandre BASTOS[1], Theodor HACK[3], Mikhail ZHELUDKEVICH[4], Mário FERREIRA[1]

[1] CICECO, DEMaC, University of Aveiro, Portugal
[2] Institute of Chemistry, University of Tartu, Estonia
[3] Airbus Central R&T, Germany
[4] Magnesium Innovation Centre−MagIC, Helmholtz-Zentrum Geesthacht (HZG), Germany

The combination of different materials in the same structure or equipment is common in modern engineering. With it comes the risk of galvanic action between those materials. A recent example is the combination of light alloys and carbon fiber reinforced plastics (CFRP) in new cars and aircrafts. The galvanic coupling of the conductive composites with the metal parts is still not well characterized and the prospect of galvanic corrosion is one of the main concerns of the transport companies. Another galvanic coupling of prime technological importance is that between zinc and iron as in galvanized steel. Often the corrosion of the zinc layer can be decreased, extending the service life and still keeping the protective galvanic action, when the ferrous substrate becomes exposed.

The addition of corrosion inhibitors, either in surface layers, paints or the environment in contact with the structure, is a traditional method of corrosion control. Due to the increasing presence of various materials in the same structure, the association of different inhibitors is becoming more common, because it is expected that the efficiency of an inhibitor changes with the material.

The aim of this work is to get an insight on the inhibition mechanisms of mixtures of cathodic and anodic inhibitors (rare earth ions and triazole derivatives, respectively) towards the corrosion of Al-CRFP and Zn-Fe couples. For such purpose, these mixtures are intended to affect both electrodes in tandem.

The work was conducted using ZRA (Zero Resistance Ammeter) measurements with a split-cell technique that enabled to separate the inhibition effect over anodes and cathodes. For the split-cell technique, cells were designed and 3D-printed, allowing for a quick, versatile and cheap cell production.

It has been found that, in some cases, the majority of the synergistic effect is given by the cathodes exposure to the mixture, and not when each electrode is exposed to the corresponding inhibitor.

This work was supported by project by: MULTISURF, this project has received funding from the European Union's Horizon 2020 research and innovation programme under the Marie Skłodowska-Curie grant agreement No 645676. The EU FP7 project PIAPP-GA-2013-612415 (PROAIR) and FCT Exploratory project IF/00856/2013/CP1162/CT0019 are also acknowledged for financial support.

Corrosion inhibition of 8-nitroquinoline for AA5052 aluminium alloy in 3wt.% NaCl solution

Daquan Zhang*,1, Hanxue Yang[1] Yaping Chen[2], Lixin Gao[1]

[1] *Shanghai Key Laboratory of Materials Protection and Advanced Materials in Electric Power, School of Environmental and Chemical Engineering, Shanghai University of Electric Power, Shanghai 200090, P. R. China*

[2] *Department of Chemical and Textile Engineering, Jiangyin Polytechnic College, Jiangshu 214400, P. R. China*

Abstract: The corrosion inhibition of 8-nitroquinoline (8-NQ) for AA5052 aluminium alloy in 3wt.% NaCl solution was evaluated by weight loss test, electrochemical impedance spectra (EIS), potentiodynamic polarisation curves, SEM/EDS surface analysis and quantum chemical calculation. The results show that 8-NQ displays a relatively high inhibition efficiency of 93.2% at 0.02 mM. 8-NQ acts as an anodic inhibitor and retards the anodic corrosion reaction. The adsorption of 8-NQ on AA5052 alloy surface is a spontaneous process and obeys Langmuir adsorption isotherm. The quantum chemical calculation was used to investigate the adsorption mode and corrosion protection mechanism of 8-NQ.

Key Words: AA5052 alloys; Corrosion; Thermal analysis; Molecular dynamics simulations; DFT

Introduction

Aluminium is the most abundant metal in the earth's crust. Aluminium and its alloy are attractive in modern industry due to their high conductivity, low density and good mechanical property [1] and [2]. AA5052 is a non-heat treatable Al–Mg alloy and widely used in automobile manufacture as the inner body panels, and in the low temperature multiple effect desalination plant as heat-exchange tube instead of copper, etc [3] and [4]. Generally, aluminium alloy has a natural corrosion protection owing to a thin surface layer of aluminium oxide on its surface [5]. However, the oxide film is easy to be broken in an environment containing corrosive ions, such as Cl^- [6]. This will lead to corrosion of AA5052 alloy. Improving the corrosion resistance of AA5052 alloy has being an important role in extending its service life. Introducing inhibitors is a common method to retard corrosion of metals due to no need of special equipment, inexpensive price and feasibility of operation. Compared with inorganic corrosion inhibitors, organic corrosion inhibitors have advantages of easy structure modification and less dosage. Most of organic corrosion inhibitors contain some polar atoms, such as O, S, N and P. They have the ability of accept or donate electrons so as to be adsorbed on metallic surface through electrostatic interaction [7], [8] and [9], or interaction with the metallic surfaces via their unshared electrons. Quinoline is an attractive inhibitor owing to its environmental-friendly property. Several quinoline derivatives are studied as corrosion inhibitors for aluminium [10], [11] and [12] and copper [13] and [14]. Nitro-group is an electron-withdrawing substituent with passivation and sterilization effect. Our recent work showed 8-nitroquinoline (8-NQ) had a better corrosion inhibition for AA5052 alloy compared with 8-aminoquinoline (8-AQ) [15]. However, there is no clear research conducted on corrosion inhibition mechanism of 8-nitroquinoline (8-NQ, Fig. 1).

In this paper, we describe our findings about the corrosion inhibition of 8-nitroquinoline (8-NQ) on AA5052 alloy in 3wt.% NaCl solution. Weight loss test, potentiodynamic

* Corresponding author, Tel./Fax: +86 21 3530 3891 e-mail: zhangdaquan@shiep.edu.cn

polarisation curves and electrochemical impedance spectroscopy (EIS) were employed to evaluate its inhibition effect. The adsorption behaviour was also studied to determine the adsorption isotherm and adsorption thermodynamic data. The surface morphology and composition were investigated by the scanning electron microscopy (SEM) coupled with the energy dispersive X-ray spectrum (EDS). Attempts to elucidate the inhibition mechanism were done by DFT (Density Functional Theory) quantum chemical calculation.

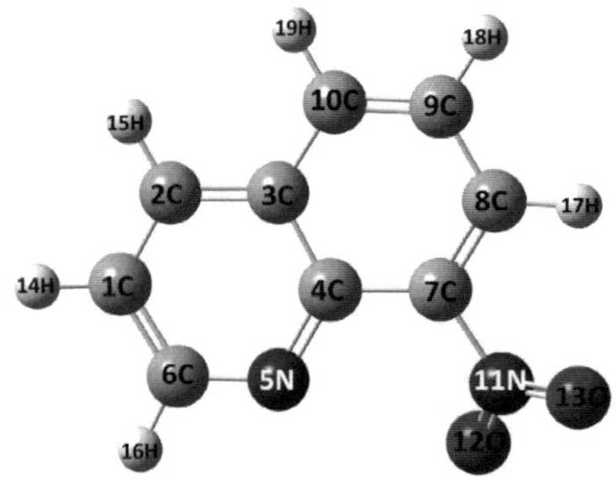

Fig. 1. Molecular structure of 8-NQ.

Experimental

Materials

The AA5052 alloy specimen have following nominal composition (wt.%): Si 0.06%, Fe 0.27%, Cu 0.10%, Mn 0.06%, Mg 2.46%, Cr 0.19%, Zn 0.10%, and Al balance. The corrosion test medium was 3wt.% NaCl solution. All chemicals and solvents were of analytical reagents and purchased from Sinopharm Chemical Reagent Co., Ltd. The text solution was prepared by the double-distilled water.

Weight loss test

Weight loss test was carried out in a thermostat water bath. The suspended coupons with the size of 50 mm×30 mm×2 mm were grinded with emery paper (01 #, 04 #, and 06 #), degreased with alcohol, washed with double-distilled water, dried and weighted accurately. These coupons were then fully immersed in 250 mL 3wt.% NaCl aqueous solution. After 6 h immersion, the coupons were erased to remove its loose surface corrosion products, re-washed and dried and re-weighted accurately to determine the weight loss. Three specimens were taken for each data to ensure their reproducibility and the accuracy of the method was within 5%. These deviations were shown in Table 1. The corrosion rate (v) is calculated by the equation [16]:

$$v(\%) = \frac{m_0 - m_{corr}}{At} \times 100 \qquad (1)$$

where m_0 and m_{corr} are specimens weight (g) before and after the immersion test, A is the exposed area (m^2), and t is the immersion time (h). The corrosion inhibition efficiency (η_{WL}) can be calculated as [16]:

$$\eta_{WL}(\%) = \frac{v_0 - v_i}{v_0} \times 100 \qquad (2)$$

where v_0 and v_i are the corrosion rate of AA5052 alloy without and with inhibitor, respectively.

Electrochemical measurements

A conventional three-electrode system was performed for electrochemical experiments using an electrochemical workstation (Solartron 1287 Electrochemical Interface with a 1260 Impedance/Gain-Phase Analyzer). The system was open to the air and the test solution was not stirred or deaerated. The Zplot and CorrWare software packages were used for data analysis and fitting. The AA5052 alloy was used as the working electrode (WE), a platinum electrode as the auxiliary electrode, and a saturated calomel electrode (SCE) as the reference electrode. The WE was sealed with epoxy resin to allow the cross section of 0.5 cm^2 to expose to the solution. Polarisation curves were obtained at a scan rate of 1 mV/s from -300 ~ +200 mV versus open circuit potential (OCP). Electrochemical impedance spectroscopy (EIS) was done at the open-circuit potential in the frequency range from 0.01 Hz to 100 kHz with 10 points per decade. The electrode system is interfered by sinusoidal alternating current with a potential perturbation of 5 mV in amplitude.

SEM and EDS analysis

The AA5052 alloy specimens (10 mm×10 mm× 2mm) were abraded by emery paper (grade 400-800-1200), rinsed with distilled water and ethanol. After immersed in 3wt.% NaCl solution for 24 h, the specimens were rinsed with distilled water, and then dried in vacuum. The surface microstructure of the specimens was observed on an emission scanning electron microscope (SU-1500, Hitachi Company). Meanwhile, the energy dispersive X-ray spectroscopy (EDS) was taken to analyse the element compositions of the specimen surface.

Quantum chemical calculations

In quantum chemical calculation, the Dmol3 package combined with the standard 6-311G (d, p) basis sets based on the density functional theory (DFT) was used [17] and [18]. Geometrical optimizations were carried out with the generalized gradient approximation (GGA) functional of Becke exchange plus Lee–Yang–Parr correlation (BLYP) [18]. Optimization of structural parameters was used to describe all stationary vibration frequency calculation at DFT level with all stationary points as minimum. The optimized specie was confirmed to have no imaginary frequencies. All calculations were performed with Gaussian 09 program package.

Molecular dynamics (MD) simulations were performed with Discover program in Materials Studio 6.0 software [17]. Al_2O_3 (111) plane was enlarged to fabricate an appropriate super-cell. A vacuum slab with 1 Å thickness was built above the Al_2O_3 (111) supercell with 14.10 Å×14.10 Å×14.10 Å of total 250 atoms. Then, the adsorption system was built by layer builder to place the optimized inhibitor molecule layers to Al_2O_3 (111) plane super-cell. The adsorption system was optimized using Smart Minimizer under Discover module. The MD simulation was performed under 298K, NVT ensemble, with a time step of 1.0 fs, simulation time of 500 ps and total step size of 5×10^5 step and record the path information every 5×10^3 steps.

Results and discussion

Weight loss test

Weight loss test is simple, reliable method to examine the corrosion inhibition efficiency of inhibitor. The corrosion rate values of AA5052 alloy with the different concentration of 8-NQ in 3wt.% NaCl solution at various temperatures are listed in Table 1. The inhibition efficiencies (η_{WL}) obtained from the weight loss are shown in Fig. 2.

Table 1. Corrosion rate of AA5052 alloy in 3wt.% NaCl solution containing different concentrations of 8-NQ at different temperature.

Concentration (mM)	Corrosion rate ($mg\ m^{-2}\ h^{-1}$)		
	298K	308K	318K
0	27.41±0.003	33.96±0.004	57.02±0.002
0.001	10.24±0.003	15.80±0.002	28.79±0.005
0.005	6.79±0.005	11.17±0.003	27.22±0.003
0.01	3.46±0.001	7.40±0.003	14.91±0.002
0.02	1.86±0.002	3.75±0.002	8.97±0.003

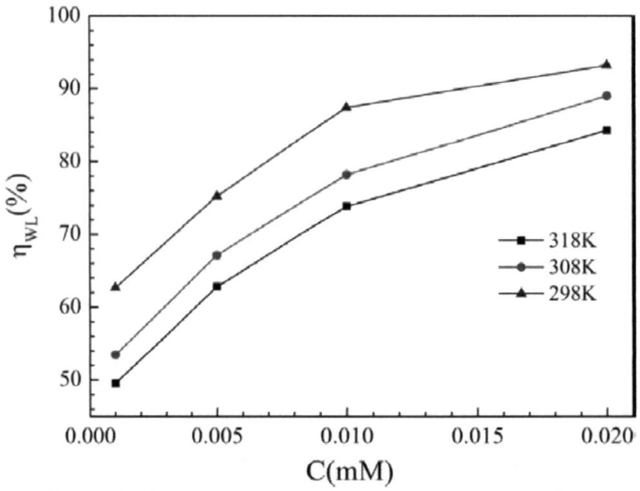

Fig. 2. Variation of inhibition efficiency (η_{WL}) vs concentration of 8-NQ for AA5052 alloy in 3wt.% NaCl solution at different temperature.

It can be seen that the corrosion rate declines sharply with the increase of 8-NQ concentration in 3wt.% NaCl solution and increases with the increase of temperature. The inhibition efficiency was 93.2% for 0.02 mM 8-NQ at 298K. An increase in temperature causes a greater corrosion damage to AA5052 alloy and a lower inhibition efficiency of 8-NQ. With the increase of the temperature, the movement of ions and molecules speeds up. It may causes the desorption of 8-NQ molecule from AA5052 alloy surface and the formation of an unstable and less protective film.

Potentiodynamic polarisation curves

Fig. 3 is the polarisation curves of AA5052 alloy after 2 h immersion in 3wt.% NaCl solution with and without 8-NQ.

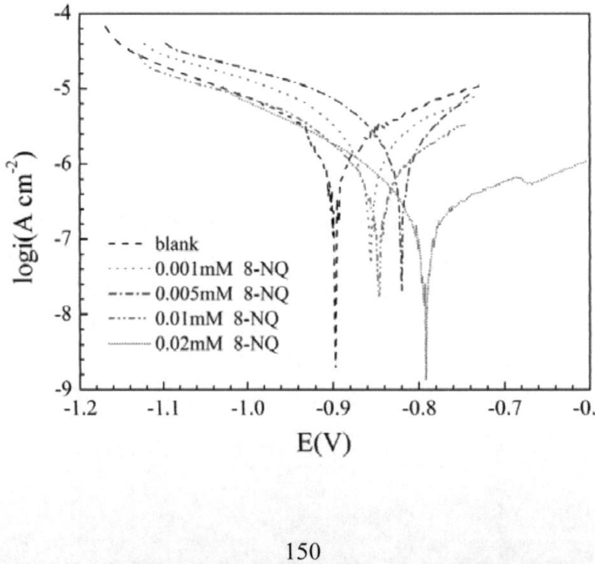

Fig. 3. Polarisation curves of the AA5052 alloy after 2 h immersion in 3wt.% NaCl solution without and with 8-NQ.

The anodic curve in the blank solution exhibits an active behaviour in the initial region. The Cl^- ions in the solution attack the defects of oxide film on the aluminium alloy surface. The anodic reaction of aluminium in 3wt.% NaCl solution is known as [19], [20] and [21]:

$$Al = Al^{3+} + 3e^- \quad (3)$$

$$Al^{3+} + 4Cl^- = AlCl_4^- \quad (4)$$

The cathodic corrosion reaction takes place as follows [19], [20] and [21]:

$$1/2O_2 + H_2O + e^- = OH_{ads.} + OH^- \quad (5)$$

$$OH_{ads.} + e^- = OH^- \quad (6)$$

Reaction (6) is the rate determining step of the cathodic corrosion process. The cathodic curves display a region of Tafel behaviour and imply that the oxygen reduction reaction is controlled by activation. As shown in Fig. 3, the corrosion potential shifts towards positive direction for AA5052 alloy electrodes and retards the anodic corrosion reaction evidently in the presence of 8-NQ. It shows that the cathodic current densities increased at low concentration of 8-NQ. This is attributed to the chemisorption of 8-NQ on the anodic and cathodic zones along the metal surface. 8-NQ can also accept exotic electrons acting as the cathodic depolarizer at low concentration. The protection by 8-NQ is strongly depended on its concentration. With the increase of its concentration, a complex film forms, which suppress the anodic reaction and retard the oxygen diffusion in the cathodic reaction. It is reported that some oxidizers can accelerate the cathodic partial reaction and retard simultaneously the anodic dissolution of the metal [22].

The polarisation parameters, such as corrosion potential (E_{corr}), corrosion current density (I_{corr}), anodic Tafel slopes (b_a), cathodic Tafel slopes (b_c), and inhibition efficiency ($\eta_I\%$), are listed in Table 2. Though extrapolating the Tafel lines to the corrosion potential can obtain the I_{corr} . The $\eta_I\%$ is calculated as [5], [23] and [24]:

$$\eta_I(\%) = \frac{I_{corr,0} - I_{corr}}{I_{corr,0}} \times 100 \quad (7)$$

where $I_{corr,0}$ and I_{corr} are the corrosion current density in without and with of inhibitor, respectively.

That current densities (I_{corr}) decrease after the addition of 8-NQ in 3wt.% NaCl solution. The corrosion inhibition is more evident with the increase of the concentration of 8-NQ, indicating 8-NQ acts as an anodic inhibitor to retard the anodic corrosion reaction of AA5052 alloy. It is observed that the anodic Tafel slope decreased in the presence of 8-NQ, indicating the 8-NQ altered the corrosion mechanism of AA5052 alloy in 3wt.% NaCl solution.

Table 2. Polarisation parameters of AA5052 alloy after 2 h immersion in 3wt.% NaCl solution at 298K without and with 8-NQ.

Concentration (mM)	b_a (mV dec^{-1})	b_c (mV dec^{-1})	E_{corr} (V vs. SCE)	I_{corr} (μA cm^{-2})	$\eta_I(\%)$
0	407.6	284.9	-0.890	8.2	-
0.001	201.6	128.2	-0.856	2.3	72.0
0.005	195.7	171.1	-0.821	2.0	75.6
0.01	190.9	150.7	-0.846	1.0	87.4
0.02	118.8	120.3	-0.795	0.6	93.2

Electrochemical impedance spectroscopy

Electrochemical impedance spectroscopy (EIS) is a non-destructive technique used to describe the characteristics of the film formed on metal surface. Fig. 4 shows the EIS diagrams of AA5052 alloy after 2 h immersion in 3wt.% NaCl solution at 298K. A single capacitive loop is measured for AA5052 alloy electrode in 3wt.% NaCl solution. The non-perfect semi-circles in Nyquist diagrams are usually because of frequency dispersion, surface roughness and heterogeneity [25]. The capacitive loop is corresponding to the interfacial reactions, especially the oxidation reactions of aluminium at the metal/oxide/electrolyte interface [26]. The corrosion reaction of AA5052 alloy is controlled by charge-transfer process, involving in the formation of Al^{3+} and OH^- ions at the metal/oxide interface. Two capacitive loops appeared in the Nyquist diagram for AA5052 alloy in the presence of 8-NQ. The first capacitive loop at high-medium frequencies is generally associated with the relaxation process in the aluminium oxide (or hydrated oxide) film presents on the Al surface [27]. The second loop at low frequencies is assigned to the aluminium dissolution [27].

Equivalent circuits in Fig. 5 were used to analyse the impedance spectra. In the circuits, R_s represents the solution resistance, R_{ct} the charge transfer resistance, R_f the inhibitive film resistance. Q_f and Q_{dl} are the constant phase elements, which represent the capacitance of the inhibitive film and the double layer capacitance, respectively. The admittance of a Q is described as [15]:

$$Y_Q = Y_0(jw)^n \quad (8)$$

where j is the imaginary root, w is the angular frequency, Y_0 is the magnitude, and n is the exponential term.

The parameter n is related to different physical phenomena containing surface in homogeneousness resulting from surface roughness, inhibitor adsorption, porous layer formation, etc. It is easily verified that the Q shows purely resistive, capacitive or inductive behaviour when n=0, 1 or -1, respectively. So the idealised capacitance values can be calculated from Q parameter values Y_0 and n using the following equation [28]:

$$C_{id} = \frac{Y_0\omega^{n-1}}{\sin(n\frac{\pi}{2})} \quad (9)$$

where ω represents the angular frequency at which imaginary value reaches a maximum on the Nyquist plots. The electrochemical parameters, according to Nyquist plots, are calculated and showed in Table 3. The chi-square (χ^2) is used to evaluate the precision of the fitted data. Table 3 reveals that the low values of χ^2 indicate that the fitted data have good agreement with the experimental data. R_{ct} is a measure standard of the corrosion rate of the metal. The inhibition efficiency (η_{EIS}) can be calculated using charge transfer resistance (R_{ct}) as follows [25]:

$$\eta_{EIS}(\%) = (1 - \frac{R_{ct}}{R'_{ct}}) \times 100 \quad (10)$$

where, R'_{ct} and R_{ct} refer to the charge transfer resistance without and with 8-NQ, respectively.

From Table 3, the R_{ct} increases with the increase of the concentration of 8-NQ, reaching its maximum value of 11.02 kΩ cm^2 at 0.02 mM 8-NQ. The C_f decreases with the increase of 8-NQ concentration. It is attributed to the formation of a thicker inhibitive film at the relatively high 8-NQ concentration. The adsorption of 8-NQ on the surface of AA5052 alloy forms a protective film and prevents Cl^- from attacking the defects of aluminium alloy.

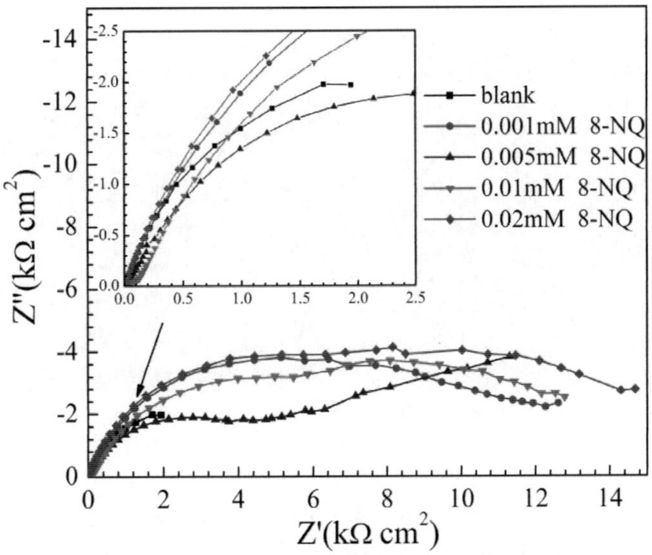

Fig. 4. Nyquist plots of AA5052 alloy after 2 h immersion in 3wt.% NaCl solution at 298K without and with 8-NQ.

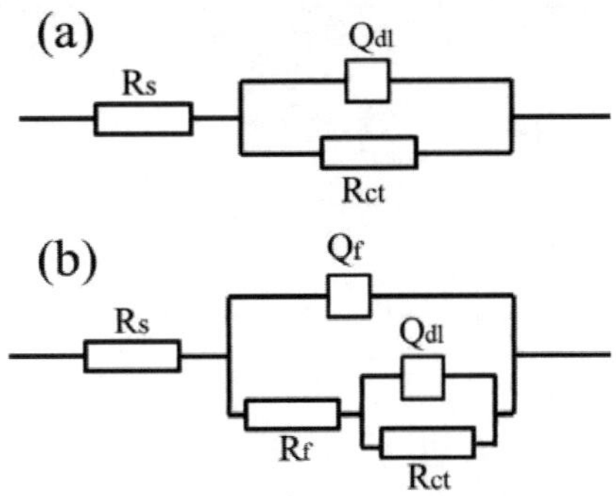

Fig. 5. Equivalent circuits of the electrochemical impedance spectra: (a) without and (b) with 8-NQ.

Table 3. Impedance parameters for AA5052 alloy in 3wt.% NaCl solution at 298K without and with 8-NQ.

Concentration (mM)	R_s (Ω cm^2)	C_f (μF cm^{-2})	R_f (Ω cm^2)	C_{dl} (μF cm^{-2})	R_{ct} (kΩ cm^2)	χ^2	η_{EIS} (%)
0	10.4	-	-	13.4	1.043	3.09×10^{-3}	-
0.001	9.4	23.3	350.2	15.2	6.272	4.26×10^{-4}	83.4
0.005	13.2	22.4	480.6	14.0	7.857	3.97×10^{-3}	86.7
0.01	21.1	13.6	632.4	17.9	10.090	1.17×10^{-3}	89.7
0.02	9.1	8.7	759.7	15.6	11.020	5.13×10^{-4}	90.5

SEM and EDS analysis

The SEM images of AA5052 alloy after 24 h immersion in 3wt.% NaCl solution without and with 8-NQ are given in Fig. 6(a) and Fig. 6(b), respectively.

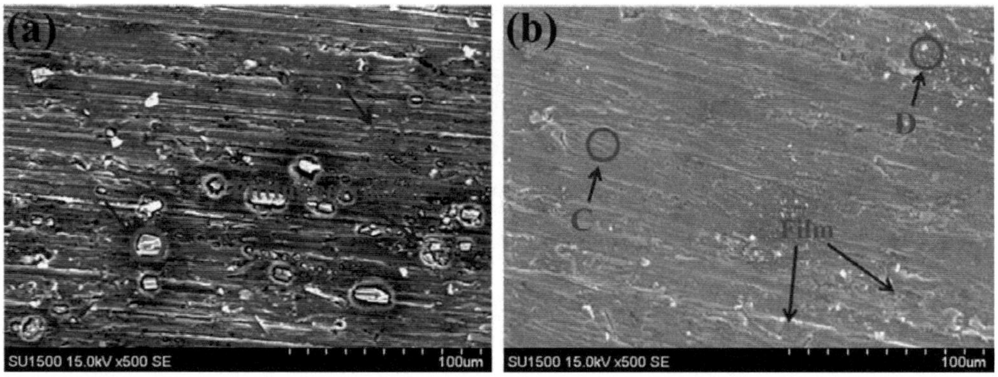

Fig. 6. SEM images of AA5052 alloy surface after 24 h immersion in 3wt.% NaCl solution at 298K without (a) and with 0.02 mM 8-NQ (b).

Table 4. Chemical composition in different area of AA5052 alloy after 24 h immersion in 3wt.% NaCl solution at 298K obtained by the EDS analysis.

Element	Composition (%)			
(wt.%)	"A" area	"B" area	"C" area	"D" area
Al	94.40	92.75	87.61	82.96
O	2.52	4.86	7.18	8.60
Mg	3.05	2.39	2.12	1.97
Fe	-	-	-	0.62
N	-	-	0.51	0.85
C	-	-	2.58	5.00

From Fig. 6(a), the pitting corrosion is obviously observed and the surface of AA5052 alloy is uneven after 24 h immersing in the 3wt.% NaCl solution. The "A" and "B" spot in Fig. 6(a) suggest an oxidation product of aluminium for its contain Al, Mg, O elements. As shown in Fig. 6(b), an obvious smooth surface of the AA5052 alloy is observed and the corrosion of AA5052 alloy is remarkably prevented by the 8-NQ inhibitor. Element N and C were detected by EDS in "C" and "D" area, confirming that the 8-NQ was adsorbed on the surface of AA5052 alloy. E.M. Sherif et al [21] and [29]. have reported that some organic inhibitors retarded the corrosion of aluminium because of their interaction with Al atoms.

Inhibition mechanism of 8-NQ inhibitor

Adsorption characteristics

Using adsorption isotherms model can describe the interaction of inhibitor and metal. Adsorption of inhibitor molecules on surface is a substitution of the adsorbed water molecules by the organic molecules. Classical Langmuir isotherm equation can better describe the monolayer adsorption behaviour [30];

$$\frac{C_{inh}}{\theta} = \frac{1}{K_{ads}} + C_{inh} \qquad (11)$$

where C_{inh} is the concentration of 8-NQ, K_{ads} is the adsorption equilibrium constant, θ is the surface coverage ratio. θ can be calculated by the following:

$$\theta = \eta_{WL} \div 100 \qquad (12)$$

Experimental results of C_{inh}/θ vs. C_{inh} yield straight lines as shown in Fig. 7, which are consistent with Langmuir's isotherm. It indicated that that monolayer adsorption reaction stands for 8-NQ on AA5052 alloy surface.

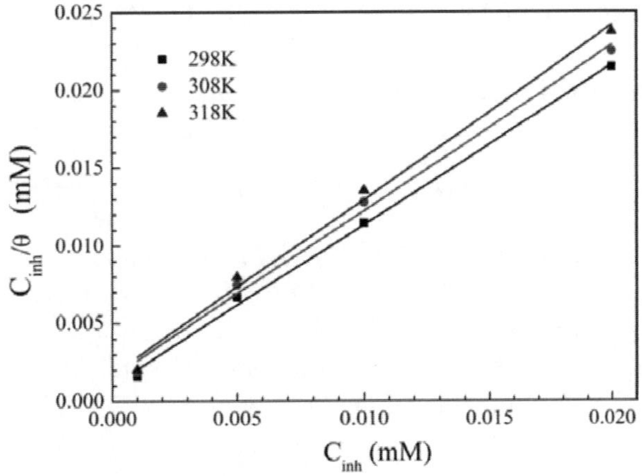

Fig. 7. Langmuir adsorption isotherm derived from gravimetric tests in 3wt.% NaCl solution.

Thermodynamic parameters

It is well-known that the adsorptive equilibrium constant (K_{ads}) represents the interaction strength between adsorbate and adsorbent. Large values of K_{ads} indicate more efficient adsorption and better inhibition efficiency. The calculated values of the adsorptive equilibrium constant (K) are listed in Table 5. K_{ads} is related to standard free adsorption heat (ΔH^{θ}_{ads}) and standard free energy of adsorption (ΔG^{θ}_{ads}), which leads to standard free entropy as [31], [32] and [33]:

$$\ln K_{ads} = \frac{-\Delta H^{\theta}_{ads}}{RT} + constant \qquad (13)$$

$$\Delta G^{\theta}_{ads} = -RT \ln 55.5 K_{ads} \qquad (14)$$

$$\Delta S^{\theta}_{ads} = \frac{(\Delta H^{\theta}_{ads} - \Delta G^{\theta}_{ads})}{T} \qquad (15)$$

where R is the gas constant (8.314 J K^{-1} mol^{-1}), T is the absolute temperature (K) and the value 55.5 is the concentration of water in solution expressed in mM. The calculated values of ΔG^{θ}_{ads} and ΔS^{θ}_{ads} are given in Table 5.

Table 5. Thermodynamic parameters for the adsorption of 8-NQ on AA5052 alloy surface.

Temperature(K)	K_{ads}	R^2	ΔH^{θ}_{ads} (kJ mol^{-1})	ΔG^{θ}_{ads} (kJ mol^{-1})	ΔS^{θ}_{ads} (J mol^{-1} K^{-1})
298	990.1	0.9967	-11.0	-27.0	53.8
308	641.0	0.9911	-11.0	-26.8	51.4
318	581.4	0.9905	-11.0	-27.4	51.7

The negative values of ΔH^{θ}_{ads} shows that adsorption of 8-NQ is an exothermic process, indicating that increasing temperature resulted in desorption of 8-NQ from AA5052 alloy surface. The negative values of ΔG^{θ}_{ads} imply that the adsorption of 8-NQ on AA5052 surface is a spontaneous process. Generally, values of $\Delta G^{\theta}_{ads} \leq -20$ kJ mol^{-1} signify the electrostatic interaction between the charged molecules and the charged metal (physisorption) and values more negative than -40 kJ mol^{-1} refer to a sharing or transfer of electrons from the inhibitor molecules to the alloy (chemisorption) [34]. The calculated ΔG^{0}_{ads} is in the range from -20 to -40 kJ mol^{-1}, indicating that both physical adsorption and chemical adsorption (mixed adsorption) occurred for 8-NQ on AA5052 alloy surface. Table 5 also reveals ΔS^{θ}_{ads} is positive. This phenomenon may be explained by the dynamic equilibrium process of sorption and desorption of organic inhibitor molecules [35], which caused the randomness going up in a system.

According to Arrhenius formula, the natural logarithm of the corrosion rate is a linear function with 1/T [36]:

$$\ln v = \frac{-E_a}{RT} + \ln A \qquad (16)$$

where v is corrosion rate (in g m^{-2} h^{-1}), E_a is the apparent activation energy, T is thermodynamic temperature (in K) and A refers to the frequency factor. By plotting ln$v$ versus 1/T, straight line graphs were obtained in Fig. 8, which were almost linear with correlation coefficients (R^2) closing to 1. The calculated values of E_a and A are given in Table 6. It is clear that both the value of E_a and the value of A increased with the increase of the 8-NQ concentration. According to reports [37], [38], [39] and [40], the value of E_a can always be used as evaluation index of corrosion effect. As a result, 8-NQ effectively retards corrosion reaction of AA5052 alloy and the corrosion rate of AA5052 alloy decreased with increasing the 8-NQ concentration.

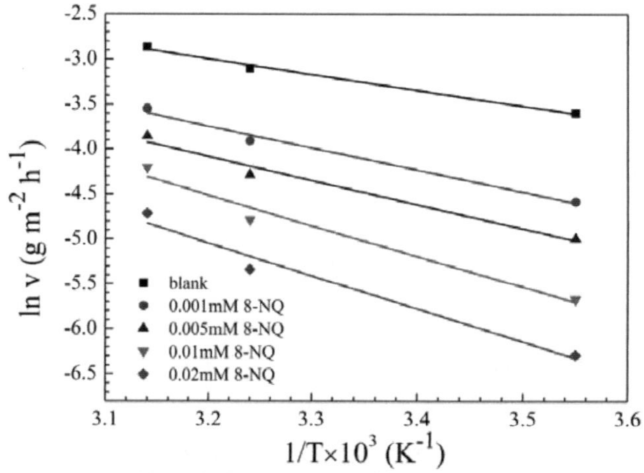

Fig. 8. Variation of lnv vs 1/T.

Table 6. E_a and A for corrosion reaction in 3wt.% NaCl solution without and with 8-NQ at different temperature.

Concentration (mM)	E_a (kJ mol^{-1})	A (g m^{-2} h^{-1})	R^2
0	14.5	13.1	0.9841
0.001	20.3	58.0	0.9741
0.005	22.1	84.2	0.9561
0.01	28.2	572.5	0.9435
0.02	30.4	781.0	0.9435

Quantum chemical calculation

Quantum chemical calculation can obtain the structural parameters of inhibitor molecule. Theoretically, the inhibitive mechanism can be explained by the chemical reactivity of the compound directly. Mulliken atomic charge calculation plays an important role in the analysis of molecular electronic structure. Generally, the more negative charge of the atom has, the better the action is as an electronic donor and the more likely it is to become active adsorptive atom [41]. Table 7 gives Mulliken charges of the C, N, O atoms. The more negative charge of atoms is found in 5N and 12O atoms, which are the electron-donating centres. This result indicates that the adsorption of 8-NQ on AA5052 alloy surface would take place through N atom in quinoline ring and O atom in nitro-group.

The dipole moment (μ) of 8-NQ is 6.38 Debye (21.2838×10^{-30} Cm), which is higher than that of H_2O (6.23×10^{-30} Cm). Generally, the higher the value of dipole moment is, the easier adsorption of molecules on the metal surface is [42]. So the adsorption of 8-NQ molecules from the aqueous solution can be regarded as a quasi-substitution process between the 8-NQ molecules in the aqueous phase [8-NQ(sol)] and water molecules at the AA5052 alloy electrode surface [H_2O(ads)]. The 8-NQ inhibitor adsorbs on AA5052 alloy surface by replacing water molecules as follows [43] and [44]:

$$Org_{(sol)} + \chi H_2O_{(ads)} \longleftrightarrow Org_{(ads)} + \chi H_2O_{(sol)}$$

which, χ is the number of water molecules replaced by an organic molecule. That is to say, 8-nitroquinoline molecules may form a kind of hydrophobic film in contact with the metal surface, thereby retard corrosion. This result is agreement with the thermodynamic calculation and the adsorption characteristic results.

In addition, the total reactivity of 8-NQ molecule is related to the distribution of the frontier molecular orbitals which includes the highest occupied molecular orbital (HOMO) and the lowest unoccupied molecular orbital (LUMO). Fig. 9 shows that the electron density distribution of HOMO and LUMO for 8-NQ. It is found that the electron density of the frontier orbital is well proportioned. The distribution of the HOMO with negative phase in 8-NQ is consistent with the charge accumulation regions of the adsorption system. The N atom in quinoline ring and O atom in nitro-group have larger electron cloud overlap, which is beneficial to form stronger interaction with Al surface [42].

Table 7. Mulliken atomic charges of 8-NQ.

atom	1C	2C	3C	4C	5N	6C	7C	8C	9C	10C	11N	12O	13O
Mulliken charge	-0.238	0.129	0.006	0.270	-0.591	0.083	-0.169	-0.169	-0.183	-0.182	0.064	-0.299	-0.263

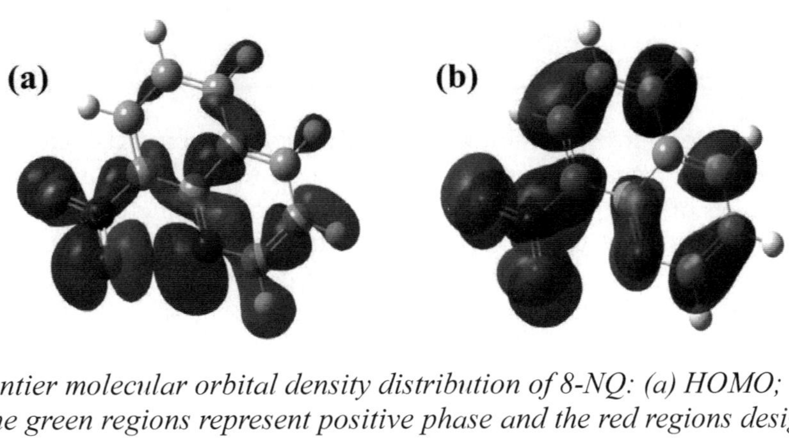

Fig. 9. The frontier molecular orbital density distribution of 8-NQ: (a) HOMO; (b) LUMO (The green regions represent positive phase and the red regions designate negative phase).

MD simulation

Molecular dynamics (MD) simulations can provide the adsorption mode of the inhibitor molecules on the metal surface at molecular level [45] and [46]. The adsorption energy ($E_{adsorption}$) between the organic inhibitor and metal surface calculated by Eq (10) [47] and [48]:

$$E_{adsorption} = -E_{interaction} = -[E_{total} - (E_{surface} + E_{organ})] \quad (17)$$

where, $E_{adsorption}$ is the adsorption energy of 8-NQ on the Al_2O_3 surface, $E_{interaction}$ the interaction energy, E_{total} the total binding energy of Al_2O_3 surface adsorbed 8-NQ, E_{organ} the binding energy of 8-NQ molecular energy before absorbed, and $E_{surface}$ the binding energy of Al_2O_3 surface.

The optimal adsorption model of 8-NQ molecules on the Al_2O_3 surface is shown in Fig. 10. The corresponding molecular dynamics simulation parameters are listed in Table 8. From Fig. 10, the 8-NQ molecule is inclined to adsorb on the Al_2O_3 surface by N atom in quinolone ring and O atom in nitro-group. We can see from Table 8 that E_{ads} of 8-NQ molecule is larger than that of H_2O molecule (26.04 kcal mol^{-1}), which indicates that when 8-NQ molecule adsorbed on the AA5052 surface, inhibitor molecules supplanted the original adsorbed water molecules and increased the resistance of the corrosion electrochemical reaction. This result is in good agreement with the dipole moment (μ) anaylsis.

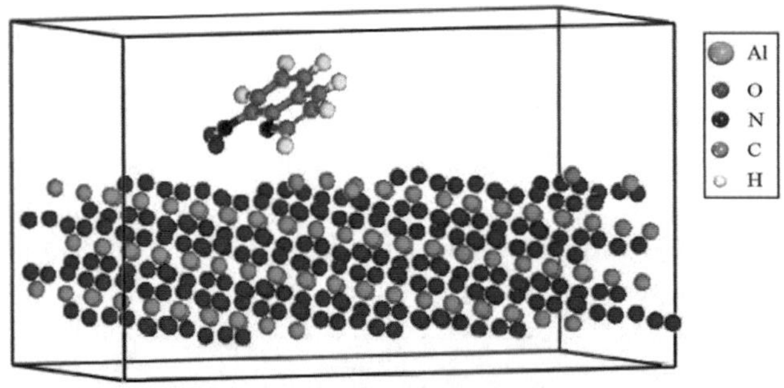

Fig. 10. Equilibrium adsorption configuration of 8-NQ on Al_2O_3 (111) planes obtained by MD simulations.

Table 8. Molecular dynamics simulation parameters of Al_2O_3 surface and 8-NQ Adsorption energy of 8-NQ on Al_2O_3 (1 1 1) surface (kJ mol^{-1}).

Inhibitor	E_{total} (kcal mol^{-1})	$E_{surface}$ (kcal mol^{-1})	E_{organ} (kcal mol^{-1})	$E_{interaction}$ (kcal mol^{-1})	$E_{adsorption}$ (kcal mol^{-1})
8-NQ	-235723.56	-235773.85	91.33	-41.04	41.04

The performance of corrosion inhibitors depends not only on the degree of coverage of the inhibitor molecules on the metal surface, but also on the stability of the protective film. The inhibition process of 8-NQ in the tested environment can be interpreted by the adsorption of 8-NQ on AA5052 alloy surface. The 8-NQ constitutes of N and O atoms in its functional groups (N-C, N-O), which have larger electron cloud density to form a strong electrostatic interaction between 8-NQ and AA5052 alloy surface [43] and [49]. That is to say, the adsorption of 8-NQ molecules may forms a protective film on the AA5052 surface, thereby retarding its corrosion effectively.

Conclusions

In the paper, the performance of the corrosion inhibitor of 8-NQ for AA5052 alloy in 3wt.% NaCl solution is investigated. The weight loss and EIS results show that the inhibition efficiency increases with the increase of concentration of 8-NQ and decreases with the increase in temperature. The maximum observed efficiency is 93.2% at 0.02 mM 8-NQ under 298K temperature. The polarization curves indicate that 8-NQ acts as an anodic inhibitor to retard the anodic reaction. The adsorption of 8-NQ on AA5052 alloy surface obeys Langmuir adsorption isotherm model and is a spontaneous and exothermic process. Quantum chemical calculation reveals that N atom in quinolone ring and O atom in nitro-group are the adsorptive centres. 8-NQ molecules adsorb on the Al_2O_3 (111) surface in the slope manner. All the data confirm the good inhibition tendency of 8-NQ, which forms a thicker film on the AA5052 alloy surface.

Acknowledgements

The research was supported by NSFC project (21776172). We are grateful to the grant from the Science and Technology Commission of Shanghai Municipality (18DZ2204400).

References

[1] A.S. Fouda, A.A. Al-Sarawy, F.S. Ahmed, H.M. El-Abbasy, Corros. Sci. 51 (2009) 485–492.
[2] S. Das, N.S. Lim, J.B. Seol, H.W. Kim, C.G. Park, Mater Design 31 (2010) 1633-1638.
[3] S.W. Song, B.C. Kim, T.J. Yoon, N.K. Kim, I.B. Kim, C.Y. Kang, Mater Trans 51 (2010) 1319-1325.
[4] A. Sommers, Q. Wang, X. Han, C. T'Joen, Y. Park, A. Jacobi, Appl Therm Eng 30 (2010) 1277-1291.
[5] H. Allachi, F. Chaouket, K. Draoui, J. Alloys Comp. 475 (2009) 300-303.
[6] B. Zaid, N. Maddache, D. Saidi, N. Souami, N. Bacha, A.S. Ahmed, J. Alloys Comp. 629 (2015) 188-196.
[7] S. Srinivasan, A. Veawab, A. Aroonwilas, Energy Procedia 37 (2013) 890-895.
[8] S. Şafak, B. Duran, A. Yurt, G. Türkoğlu, Corros. Sci. 54 (2012) 251-259.
[9] B. Obot, N.O. Obi-Egbedi, Int. J. Electrochem. Sci. 4 (2009) 1277-1288.
[10] S.V. Lamaka, M.L. Zheludkevich, K.A. Yasakau, M.F. Montemor, M.G.S. Ferreira, Electrochim. Acta 52 (2007) 7231-7247.
[11] S.L. Granese, B.M. Rosales, C. Oviedo, J.O. Zerbino, Corros. Sci. 33 (1992) 1439-1453.
[12] L.B. Tang, X.M. Li, Y.S. Si, G.N. Mu, G.H. Liu, Mater. Chem. Phys. 95 (2006) 29-38.
[13] M. Cubillos, M. Sancy, J. Pavez, E. Vargas, R. Urzua, J. Henríquez-Roman, B. Tribollet, J.H. Zagal, M.A. Páeza, Electrochim. Acta 55 (2010) 2782-2792.
[14] G.P. Cicileo, B.M. Rosales, F.E. Varela, J.R. Vilche, Corros. Sci. 40 (1998) 1915-1926.
[15] D.P. Wang, D. Yang, D.Q. Zhang, K. Li, L.X. Gao, T. Lin, Appl. Surf. Sci. 357 (2015) 2176-2183.
[16] J. Halambek, K. Berković, J. Vorkapić-Furač, Mater. Chem. Phys. 137 (2013) 788-795.
[17] T.H. Li, C.M. Wang, S.W. Yu, X.Y. Liu, X.H. Li, X.G. Xie, Chem. Phys. Lett. 463 (2008) 334-339.
[18] A.D. Becke, J. Chem. Phys. 88 (1988) 2547-2553.
[19] E.S.M. Sherif, Int. J. Electrochem. Sci 7 (2012) 4847-4859.
[20] E.S.M. Sherif, Int. J. Electrochem. Sci 6 (2011) 1479-1492.
[21] E.M. Sherif, S.M. Park, Electrochim. Acta 51 (2006) 1313-1321.
[22] N.D. Tomashov, Corrosion and protection of the metals, in (Gov. Ed.) "Nauka", Moskow, 1952, pp. 159.
[23] X.H. Li, S.D. Deng, F. Hui, Corros. Sci. 53 (2011) 302-309.
[24] D.Q. Zhang, B. Xie, L.X. Gao, H.G. Joo, K.Y. Lee, J. Appl. Electrochem. 41 (2011) 491-498.
[25] A. Singh, Y.H. Lin, W.Y. Liu, S.J. Yu, J. Pan, C.Q. Ren, K.H. Deng, J. Ind. Eng. Chem. 20 (2014) 4276-4285.
[26] K.F. Khaled, M.M. Al-Qahtani, Mater. Chem. Phys. 113 (2009) 150-158.
[27] M.L. Zheludkevich, K.A. Yasakau, S.K. Poznyak, M.G.S. Ferreira, Corros. Sci. 47 (2005) 3368-3383.

[28] M. Hosseini, S.F.L. Mertens, M.R. Arshadi, Corros. Sci. 45 (2003) 1473-1489.
[29] E.M. Sherif, F.H. Latief, H. Junaedi, A.A. Almajid, Int. J. Electrochem. Sci 7 (2012) 4352-4361.
[30] P.B. Raja, A.K. Qureshi, A.A. Rahim, H. Osman, K. Awang, Corros. Sci. 69 (2013) 292-301.
[31] A.Y. Musa, Corros. Sci. 52 (2010) 3331-3340.
[32] G.N. Mu, X.H. Li, Q. Qu, J. Zhou, Corros. Sci. 48 (2006) 445-459.
[33] T.P. Zhao, G.N. Mu, Corros. Sci. 41 (1999) 1937-1944.
[34] X.H. Li, S.D. Deng, F. Hui, Corros. Sci. 53 (2011) 3704-3711.
[35] S.D. Deng, X.H. Li, X.G. Xie, Corros. Sci. 80 (2014) 276-289.
[36] Q.J. Xu, Z.Y. Wan, G.D. Zhou, R.H. Yin, W.M. Cao, C.J. Lin, Chin. J. Chem. 26 (2008) 1579-1584.
[37] X.H. Li, S.D. Deng, H. Fu, Corros. Sci. 62 (2012) 163–175.
[38] J.M. Falcón, F.F. Batista, I.V. Aoki, Electrochim. Acta 124 (2014) 109-118.
[39] M.M. Osman, R.A. El-Ghazawy, A.M. Al-Sabagh, Mater. Chem. Phys. 80 (2003) 55-62.
[40] A.S. Algaber, E.M. El-Nemma, M.M. Saleh, Mater. Chem. Phys. 86 (2004) 26-32.
[41] S. Martinez, I. Stern, Appl. Surf. Sci. 199 (2002) 83-89.
[42] X.H. Li, S.D. Deng, X.G. Xie, Corros. Sci. 81 (2014) 162-175.
[43] B.G. Ateya, B.E. El-Anadouli, F.M. El-Nizamy, Corros. Sci. 24 (1984) 509-515.
[44] M. Şahin, S. Bilgic, H. Yılmaz, Appl. Surf. Sci. 2002, 195, 1-7.
[45] K. Ramji, D.R. Cairns, S. Rajeswari, Appl. Surf. Sci. 254 (2008) 4483-4493.
[46] F. Zhang, Y.M. Tang, Z.Y. Cao, W.H. Jing, Z.L. Wu, Y.Z. Chen, Corros. Sci. 61 (2012) 1–9.
[47] Y.G. Yan, X. Wang, Y. Zhang, P. Wang, X.H. Cao, J. Zhang, Corros. Sci. 73 (2013) 123–129.
[48] J. Zhang, W.M. Zhao, W.Y. Guo, Y. Wang, Z.P. Li, Acta Phys. Chim. Sin. 24 (2008) 1239–1244.
[49] B. Xu, Y. Liu, X.S. Yin, W.Z. Yang, Y.Z. Chen, Corros. Sci. 74 (2013) 206–213.

Oxidation behavior and mechanism of 5 wt.% Cr ferritic steel in air

Enhui WANG[1], Yafeng CHEN[1], Zhi FANG[1], Kuochih CHOU[1], Xinmei HOU[1]

[1] *University of Science and Technology Beijing, China PR*

Low-Cr ferritic steels combining a low price and good properties are widely used as heat resistance structure working at high temperature. Under such conditions, the steels are easily oxidized, leading to the structure failure. Although introduction of alloying elements seems to be a feasible way to increase oxidation resistance, relevant work have been focused on high-Cr (≥9 wt%) ferritic steels. The effect of alloying elements on low-Cr steels still remains unclear. To clarify this problem, a modified low-Cr steel, i.e. 5 wt.% Cr (5Cr) ferritic steel with addition of appropriate alloying elements is chosen as the objective in our work and its oxidation behavior in air is investigated. The main results are as follows: besides continuous Cr and Si spinels ribbon, the connected CuO layer is adjacently formed in inner oxidation layer (IOL) at 500-600 °C. Their synergistic effects imped the growth of Fe-rich oxide film towards outer oxidation layer (OOL), which contributes to the oxidation resistance. With temperature increasing, the protective distribution of Cu and Si elements in oxide scales are gradually destroyed and even the spallation of oxide scale occurs at 1000 °C. Specific oxide composition indicates the oxidation is controlled by oxygen inward diffusion. Based on this, a gas-solid model is adopted to describe and predict the oxidation kinetics of 5Cr steel at 500-900 °C.

Extraction of essential oil from Artemisia herba alba and it application in the protection of Aluminum in HCl medium

Nacer HECHICHE[1]

[1] *Université Mouloud Mammeri Tizi-Ouzou, Algeria*

Extraction of essential oil from Artemisia herba alba and it application in the protection of Aluminum in HCl medium

N.Hechiche [1],D.Boughrara[1],L.Hamadou[1], A.Kadri[1]

Université Mouloud MAMMERI de Tizi-Ouzou Laboratoire de Physique et Chimie des Matériaux (LPCM),

BP 17 RP, 15000, Algérie

Abstract

Acids play a very important role in the industry especially in (scrubbing, industrial cleaning, acid scaling, etc.). However, the action of these agents alters metals and can sometimes, in addition to cost, lead to serious accidents. The use of corrosion inhibitors to combat this phenomenon is effective, however synthetic compounds are highly toxic and most of them are banned. As a result, the use of oils and plant extracts has aroused the interest of the scientific community as a good source of "green" inhibitors. It is within this framework that our present study is taking place. Indeed, starting from a green plant in this case the white artemisia, picked in the east of Algeria , the essential oil is extracted to study its effectiveness with respect to the corrosion of pur aluminium in HCl 1mol. L^{-1}. The extraction of the oil was carried out by hydrodistillation with an efficiency of (0.9 - 1.1%), the extract obtained was then characterized by gas chromatography coupled with mass spectroscopy. The study of the electrochemical behavior of aluminum in hydrochloric medium in the presence of this extract (temporal variation of the abandonment potential, potentiodynamic polarization, electrochemical impedance spectroscopy and gravimetric measurements allowed us to demonstrate an inhibitory efficiency of 95% and the extract acts as cathodic inhibitor.

Anti-corrosion coatings produced from complex oxides

Oleg CHIZHKO[1]

[1] *German-Russian College, Russia*

Let us regard a condensation of oxide phases from liquid state of matter. We introduce the case of molten slag after combustion of conventional fuels and municipal waste. The temperature of treatment corresponds to the conditions of inflammation of natural coal at 2000°C. The ash fraction of these fossil compounds consists from aluminates and silicates. Together with synthesis of mullite, we investigate an influence of Tl_2O_3 on formation of stable phases with aim of their implementation in anti-corrosion protectors.

A liquid slag is characterized with long range order of oxy-anionic units, which are the constituents of pure oxides. In the conditions of high temperatures, the metal-oxygen components are all negative because of their affinity to attract O-atom. Thallium improves this property after diffusion in the composition of aluminum and silicon oxides. Near the melting point, the oxygen affinity becomes an excess value because the emission of free O_2 allows a formation of oxide complexes. They are additionally bonded with the internal electron levels. Now, the molten state is the mixture of oxy-cations and oxy-anions. The positive and negative constituents discharge and build sub-lattices of solid body. In the case of non-turbulent processes, thermal treatment of slow condensation is directed to the production of bigger size of material grains. This consequence is desirable because the natural pigmentation of oxides depends on their heat conductivity. Thus, the pure substances build an ordinary value of particles, but the coating of such agglomerates is problematic. The conventional solution of this task is the application of more active components, which substitute the hexagonal family of aluminates and silicates.

The addition of Y_2O_3 promotes the total stability of Al_2O_3- and Fe_2O_3-composites. They become the garnet $(Fe,Al)_5Y_3O_{12}$ and perovskite $(Fe,Al)YO_3$ structures. The system Al_2O_3-Fe_2O_3 presents the area of substitution of $FeAl_3O_6$ as the result of diffusion processes. The field of two phases diminishes together with the growth of temperature and proves an increasing of solubility of the both oxides. At 1044°C, the aluminates destroy into the orthorhombic structure of $FeAlO_3$. Stabilization of yttrium oxide is rational together with Tl_2O_3 because of their structural analogy. They are both BCC, so we investigate the reciprocal exchange of $TlYO_3$.

Oxidation behavior of Alx(CoCrFeNi)100-x high entropy alloys

Juliusz DĄBROWA[1], Grzegorz CIEŚLAK[2], Mirosław STYGAR[1], Aleksander GIL[1], Tadeusz KULIK[2], Marek DANIELEWSKI[1]

[1] *AGH University of Science and Technology, Poland*
[2] *Warsaw University of Technology, Poland*

Oxidation behavior of the different FCC and FCC+B2 structured high entropy alloys (HEAs) from Al-Co-Cr-Fe-Ni metallic system was investigated. These materials were tested under both isothermal and thermal cycling conditions in air atmosphere, at temperature of 1273 K. The results indicate relatively severe spallation in all materials. The scales exhibit multiphase structures, with Cr_2O_3 being the main product of oxidation. The microstructure of the samples after both isothermal and cyclic oxidation is compared. The early stages of oxidation were investigated for all materials, with samples being subjected to number of short-term oxidation annealings. To reduce spallation of the scales, Gd was deposited on surfaces of selected alloys. The preliminary results on Gd influence on the oxidation behavior of the HEAs are shown.

The oxidation of severe deformed FeAlCr intermetallic alloy

Marek ZAJUSZ[1], Janusz CEBULSKI[2], Dorota PASEK[2], Marek DANIELEWSKI[1]

[1] AGH University of Science and Technology, Poland
[2] Silesian University of Technology, Poland

The effect of severe plastic deformation of FeAlCr alloys on their oxidation behaviour is presented. We show the changes of metallic substrate composition caused by oxidation. The experimental investigations are supported by the numerical simulation of the selective oxidation process. Combination of both experimental and theoretical methods, allows accelerating laboratory tests, as well as estimating the lifetime of the hot extruded FeAlCr intermetallic alloy for different oxidation temperatures. It also provides a possibility for increase of materials lifetime through conscious control of the selective oxidation process. The microcrystalline structure of FeAlCr alloys obtained by extrusion increases effective diffusivities in alloy, as well as improves the corrosion resistance of materials.

The Influence of 3-Amino-5-Methylthio-1H-1,2,4-Triazole on Pitting Behavior of 304 Stainless Steel

Gozde TANSUG[1]

[1] *Cukurova University Ceyhan Engineering Faculty, Turkey*

The pitting corrosion behavior of 304 type stainless steel was studied in 3.0% NaCl solution in the presence of 3-amino-5-methylthio-1H-1,2,4-triazole. The impact of inhibitor on initiation, progress of metastable pittings and growth transition of stable pits were investigated. Potentiodynamic and potentiostatic measurements were carried out in chloride environment. The results indicated that the inhibitor is capable of shifting the pitting potential towards more noble values and decreasing passive current density. Moreover, the mechanism has been utilized to explain the impact of the inhibitor on pitting behavior.

Evaluating the Oilfield Chemical Treatment Program of a Saudi Aramco Water System

Yahya T. ALJANABI[1], Mostafa M. ALSALEM[1], Anaam H. SHAIKH ALI[1], Ali A. AL-MOHAISH[1], Mansour A. AL-SHAFEI[1], Qiwei WANG[1], Waleed N. AL-NASSER[1], Hassan A. AL-BASRI[1]

[1] *Saudi Aramco, Saudi Arabia*

A multi-disciplinary team (corrosion, compositional analysis, scale, water geochemical analysis, and microbiology sub-teams) was formed in order to evaluate the oilfield chemical treatment program of the Manifa Water System.

The objectives of this study were to:

- determine the main corrosion-related components present in the water system by using various advanced laboratory analytical techniques;
- recommend the use of appropriate chemical treatment including corrosion inhibitors, scale inhibitor, and biocides based on the field data and test results to enhance the integrity of the water injection pipelines; and
- conduct compatibility performance evaluation between current and recommended chemicals.

Following are summary of the results obtained by each sub-team and the corresponding conclusions and recommendations:

- Advanced compositional analysis: Performed a thorough characterization of collected sludge and filtrate samples. The sludge samples comprised of mixtures of inorganic and organic ingredients. The inorganic part was found to be more predominant in all samples. It comprised of various levels of Ca-rich and Fe-rich compounds, which are typical of scaling and corrosion products. The organic part was found to be a mixture of hydrocarbons, amines and sulfur-containing compounds. Traces of the corrosion inhibitor used in the downhole squeeze treatment were also detected. For the water filtrate samples, detailed analysis indicated the presence of sand particles in addition to scale and corrosion products. Considering all of the above, the sludge is comprised of corrosion and scale products along with sand particles which are amalgamated by the hydrocarbon fluids from the downhole corrosion inhibitor squeeze treatments.
- Scale analysis: Completed scaling predictions for Manifa Water Injection Flanks and based on the results recommended implementing scale inhibition and monitoring programs. The recommended monitoring technique is scale coupons. Also, suspended solids samples should be regularly collected and analysed. Recommended injecting a scale inhibitor at an initial dosage of 10 ppm.
- Water analysis: Completed geochemical analysis of Abu Hadriyah (ABHD) water samples. The analysis showed big variations in TDS and constituents levels of the waters produced from the different wells. Since these waters will eventually be mixed, worst case scenarios should be considered for conducting oilfield chemical treatment programs. The relatively high chloride, Cl^-, contents could lead to localized corrosion through breakdown of the protective $FeCO_3$ layer. Sulfates are also present which are considered good nutrients for microbes.
- Microbial analysis: Completed microbial analysis of water and sludge samples, and corrosion coupons. The results revealed high to moderate numbers of corrosion inducing microbes in water samples, high numbers in the sludge samples, and high to moderate numbers in the corrosion coupons. Recommended employing monitoring and an anti-microbial treatment programs which are being evaluated and will be implemented.
- Corrosion analysis: Both corrosion and scale products were found in the sludge samples. The source of hydrocarbons in the sludge was traced to the downhole corrosion inhibitor squeeze treatments. The corrosion inhibitors for surface and for downhole treatments were found compatible at the different conditions in terms of corrosion performance. The downhole corrosion inhibitor were found to become immiscible in water at higher concentrations. The surface corrosion inhibitor and the scale inhibitor were found compatible in terms of corrosion performance. The corrosion inhibitors were found thermally stable at 70°C. Also, recommended employing monitoring programs – including locations - for corrosion, iron and manganese contents, and dissolved oxygen.
- According to worldwide field experience, the type of corrosion found at Manifa is the most common

corrosion type in water injection pipelines and is referred to as "channelling corrosion".

- The main conclusion from a large portfolio of carbon steel sea-water injection pipelines is that almost always inadequate control (housekeeping) in terms of maintenance and chemical control lead to channelling corrosion.

Continuous Downhole Corrosion Inhibitor Treatment of Selected Saudi Aramco Water Supply Wells

Yahya T. ALJANABI[1], Mostafa M. ALSALEM[1], Mustafa S. AL-JAWAD[1], David R. LEWIS[1]

[1] *Saudi Aramco, Saudi Arabia*

Downhole corrosion at the water supply wells of Qatif and Berri fields at Saudi Aramco is managed using downhole corrosion inhibitor squeeze treatment. The main objective of this study is to replace the downhole squeeze treatment currently used with a more effective and efficient downhole continuous corrosion inhibitor injection.

Following are the conclusions of this study based on the performed laboratory assessments:

- Based on the test results of thermal stability, viscosity changes, solubility in synthetic formation water, and foaming tendency, three candidate corrosion inhibitor packages ranked highest and these are: SUM, BHU, and REO. Based on the results of TGA analysis of these three candidate inhibitors, the sequence of thermal stability starting with the more thermally stable is: SUM > BHU > REO. At 79°C, both SUM and REO were completely soluble in synthetic Wasia water up to a concentration of 10,000 ppm. The chemicals that underwent the least changes in viscosities after thermal exposures were SUM and BHU.
- Using LPR measurements conducted at Berri downhole conditions of 79°C with continuous CO_2 purging and at the three chemical concentrations 5, 10, & 15 ppm, the best three performers ranked in ascending order are: BHU > REO > SUM.
- The three highest ranked corrosion inhibitors, namely BHU, REO, & SUM, were found compatible with the corrosion inhibitor used in the surface facilities.
- Based on the average weight loss corrosion rates measured using the rotating cage method for test solutions containing 25 ppm and 100 ppm, the overall ranking based on calculated inhibition efficiencies in ascending order is: BHU > SUM > REO. These tests were conducted at the downhole temperature of Berri water supply wells, 175°F/79°C, and at a cage rotation speed of 500 RPM. The tests were performed using standard carbon steel test coupons CS1018 in synthetic Wasia water with 15 psi CO_2 gas and balance 985 psi N_2 gas.
- Among the three best performers, SUM gave the best protection against localized corrosion at 25 ppm. At 100 ppm, the results of the three corrosion inhibitors were similar. Since the inhibitor dosage could be fluctuating in the field, chemicals with the best performance against both localized and uniform corrosion at both low and high inhibitor concentrations were recommended for field trials.
- Based on previous work, the corrosion inhibitor currently used for downhole squeeze treatment, i.e. INI, is biocide enhanced. A biocide enhanced corrosion inhibitor is not recommended for continuous downhole injection since the bacteria could become immune to it with time. A separate alternating biocide treatment could be necessary if the water supply well is contaminated with bacteria.
- Three chemicals were recommended for field trials and these were: SUM, REO, and BHU. However, care should be exercised when using BHU, since it could induce localized corrosion at low concentrations.

The synergistic inhibitive effect of nitro aniline compounds and iodide ions on mild steel in 1M HCl: Experimental and theoretical approach.

Francis OJO[1], **Isaac ADEJORO**[2], **Godfrey AKPOMIE**[3], **Babatunde OGUNYEMI**[4]

[1] *Bingham University, Nigeria*
[2] *University of Ibadan, Nigeria*
[3] *University of Nsuka, Nigeria*
[4] *Federal University Otuoke, Bayelsa State, Nigeria*

Abstract

The effect of iodide ions on the inhibitive performance of ortho, meta and para nitro aniline in 1M HCl for mild steel has been studied using weight loss method (gravimetric) measurements at 303 and 333K. Results obtained show that the presence of the nitro aniline compounds in 1M HCl solution inhibits the corrosion process of mild steel. The inhibition efficiency of the nitro aniline compounds was found to increase with increase in concentration of these compounds and decrease with increase in temperature. Its adsorption was found to be physical, exothermic and spontaneous as confirmed by values of activation energy and free energy of adsorption (not up to -20 kJ mol^{-1} for free energy of adsorption and below 80 kJ mol^{-1} for activation energy) and also fitted the Langmuir adsorption model. Addition of iodide ions synergistically increased the inhibition efficiency of the nitro aniline compounds. Quantitative Structure Activity Relationship (QSAR) approach was used on a composite index of some quantum chemical parameters to characterize the inhibition performance of the studied molecules. The results showed that the Inhibition Efficiency was closely related to some of the quantum chemical parameters.

Investigation of The Friction and Wear Behavior of a 3rd Generation γ–TiAl Based Alloy

<u>Lukas MENGIS</u>[1], Mathias C. GALETZ[1]

[1] *DECHEMA e.V., Germany*

Due to their specific high-temperature strength, elastic modulus as well as their low density γ-TiAl based alloys have received increasing attention in terms of high-temperature applications [1]. In this respect, titanium aluminides have already partially substituted heavier Ni-based alloys as aerospace components such as primarily low-pressure turbine blades. As turbine blades are usually fitted into the turbine disk by a specific dovetail connection, components will be inevitably in contact with different metallic surfaces and therefore subject to friction and wear [2]. Hence, tribological properties are critically important to the overall lifetime performance of these heavily loaded components.

Within the scope of this study a basic analysis of the tribological properties of the 3rd generation TNM-B1 alloy has been performed using a pin-on-disk apparatus under atmospheric conditions. Primarily, friction and wear parameters were investigated under variation of the applied load as well as the speed of rotation sliding TNM-B1 against itself as well as certain single TiAl phases. Secondly, a variation of the counterpart material was undertaken to simulate conditions relevant for today's turbine engines. Hereby, Ni- and Co-based superalloys have been used as counterpart materials. Conclusively, characteristic wear rates were calculated and corresponding wear mechanisms are discussed to extensively depict the tribological performance of the substrate.

<u>References:</u>

[1] S. Mayer et al., "Intermetallic β-Solidifiying γ-TiAl Based Alloys - From Fundamental Research to Application", Advanced Engineering Materials, Volume 19, Issue 4, 2017

[2] K. Miyoshi et. al., "Fretting Wear of Ti-48Al-2Cr-2Nb", Tribology International, Volume 36, Issue 2, 2003, 145-153

Dandelion extract as a green corrosion inhibitor for carbon steel in 0.5 M H2SO4 solution

Mohamed DEYAB[1]

[1] *Egyptian Petroleum Research Institute, Egypt*

The inhibition effect of Dandelion extract on the corrosion of carbon steel in 0.5 M H_2SO_4 solution was investigated by chemical (mass loss), electrochemical (EIS) and scanning electron microscopy (SEM) methods.

The presence of Dandelion extract in acid solution affects on corrosion behavior of carbon steel due to the adsorption of Dandelion extract constitutes on the metal surface. The maximum inhibition efficiency (81.5%) was obtained at extract concentration of 400 mg/l. The energy barrier for the corrosion process becomes greater in the presence of Dandelion extract. SEM examinations clarified a decrease in corrosion attacks in presence of Dandelion extract.

Outward diffusion through protective Alumina scales

Torben BOLL[1], Olof BÄCKE[2], Martin HEILMAIER[1], Kinga A. UNOCIC[3], Bruce A. PINT[3], Krystyna STILLER[2]

[1] *Institut für Angewandte Materialien - Werkstoffkunde (IAM-WK), Karlsruher Institut für Technologie (KIT), Germany*

[2] *Department of Physics, Chalmers University of Technology, Sweden*

[3] *Materials Science and Technology Division, Oak Ridge National Laboratory, United States of America*

While it is established that the main process of alumina scale growth on NiAl-alloys and FeAlCr-alloys, after a protective layer has been formed, is the diffusion of oxygen through the scale to the metal, there is also a contribution of outward diffusion. The diffusion processes, which happens along grain boundaries (GBs), can be manipulated largely by the choice of alloying reactive elements segregating towards GBs. The main contribution to this outward diffusion is stemming from Al, which was already shown in previous studies [1]. For this study, we applied scanning electron microscopy, transmission electron microscopy and atom probe tomography (APT) to analyze oxidized NiAl and FeAlCr with and without addition of reactive elements like Hf, Y and Zr. Due to the high chemical and spatial resolution APT provides, it was possible to observe and even quantify GB segregation of the reactive elements and Fe, Cr or Ni. By applying a two-step oxidation process, for which after an initial oxidation the surface was slightly polished and then oxidized a second time, and subsequently performing a site specific preparation at the newly grown ridges on top of the GBs, it was even possible to quantify the outward diffusion of these metals. In the case of the NiAl-alumina the non-Al outward diffusion was four orders of magnitudes smaller than the Al-outward diffusion, which itself was six orders of magnitude smaller than the O-inward diffusion.

[1] T. Boll, K. A. Unocic, B. A. Pint, A. Mårtensson, and K. Stiller, "Grain Boundary Chemistry and Transport Through Alumina Scales on NiAl Alloys," *Oxid. Met.*, pp. 1–11, 2017.

Effects of TiO2 nanoparticles on anti-corrosion performance of brass in simulated water containing surfactant

Hong-Hua GE[1], **Kai WU**[2]

[1] *Shanghai University of Electric Power, China PR*

[2] *Shanghai University of Electric Power,, China PR*

The anti-corrosion performance of brass in simulated cooling water containing or free of surfactant sodium dodecyl benzene sulfonate (SDBS) and TiO_2 nanoparticles were analyzed by electrochemical impedance spectroscopy (EIS), scanning electron microscope (SEM), energy spectrum analysis (EDS), and X-ray diffraction (XRD). The relationship between the surface states and corrosion resistance of brass was investigated by EIS. The results showed that the surface of brass were mainly covered with zinc compound $Zn_5(OH)_6(CO_3)_2$ as corrosion product in simulated cooling water (SCW). In SCW containing SDBS, a large amount of adsorbed SDBS was found on the brass surface, and the corrosion products were mainly cuprous oxide (Cu_2O). In TiO_2 nanofluid (simulated water containing nanoparticles), the surface of the brass was relatively flat and smooth and mainly containing Cu_2O, with no obvious adhesion of the SDBS aggregates and accumulation of corrosion products. TiO_2 nanoparticles inhibited the adsorption of SDBS on the brass surface. The results of EIS showed that the brass had a larger reaction resistance in SCW containing SDBS, indicating the good protective performance of the adsorbed SDBS film on brass surface. The reaction resistance of the brass electrode was the smallest in TiO_2 nanofluid, indicating that TiO_2 nanoparticles in solution promoted the corrosion of brass.

Key Words: TiO_2 nanofluid; Sodium dodecyl benzene sulfonate; Brass; Surface state; Corrosion

Creep-oxidation behavior of cast alloy 20Cr-32Ni-1Nb and its weld joints in air

Mathias C. GALETZ[1], Ellen BERGHOF-HASSELBÄCHER[1], Ali SOLIMANI[1]

[1] *DECHEMA Forschungsinstitut, Germany*

High temperature oxidation behavior of the heat resistance cast alloy 20-32Nb (20Cr-32Ni-1Nb) during creep test in air at 816-982°C and under different stress levels between 23 and 75 MPa for targeted rupture times between 500-5000 hours were studied. The oxide scale thickness formed during the creep test was measured for different rupture times and temperatures. Along with the oxide scale thickness, Cr-depletion of the subscale substrate were monitored and correlated to the oxide growth. The kinetics of oxidation was determined using Wagner's theory of oxidation. The parabolic rate constant (k_p) was calculated for each test temperature. The corresponding data at different test temperatures provided the activation energy of oxidation (Q) in presence and absence of the external load. Microstructural evolution such as internal oxidation and compositional changes at the subscale alloy substrate were investigated and correspondingly correlated to the oxidation behavior. A similar approach was employed to determine and compare the oxidation behavior of selected weld joints such as alloy 21-33Mn (21Cr-33Ni-4.5Mn) and Alloy 617 with that of the base metal. The oxidation rate of all investigated alloys can be ranked from the slowest to the fastest as Alloy 617 < Alloy 20-32Nb < Alloy 21-33Mn, respectively.

Microstructural characterization of oxide scale on ductile iron pipes using scanning electron microscopy

Anna WASSILKOWSKA[1], Wojciech DĄBROWSKI[2]

[1] *Cracow University of Technology (CUT), Poland*

[2] *Institute of Water Supply and Environmental Monitoring (CUT), Poland*

The knowledge of the oxide scale microstructure and composition is critical in understanding the mechanisms of soil corrosion protection. In Europe, of ductile iron pipes are protected by factory external coatings including the zinc based coating [1]. Heat treatment is carried out after pipes centrifugal casting in order to obtain ferritic matrix of ductile iron with improved corrosion resistance. High-temperature treatment of ductile iron pipe causes building of double-layered porous and brittle oxide scale [2]. The mechanical splitting of the protective coating usually takes place between the above mentioned sub-layers of the oxide scale. The composition and microstructure of scale on selected commercial pipes have been studied using scanning electron microscopy (SEM). Recently energy dispersive spectroscopy (EDS) has been approved to be powerful technique for distinguishing the different layers in a scale.

Keywords:

ductile iron; centrifugal casting; annealing; zinc based coatings; scanning electron microscopy

References:

1. EN 545:2010, Ductile iron pipes, fittings, accessories and their joints for water pipes.
2. A. Wassilkowska, New generation of water supply ductile iron pipes, 2017, in Polish, ISBN 978-83-7242-946-9

A new solution for the corrosion protection in the form of functional coatings based on silicon oxycarbide glasses

Maciej BIK[1], **Piotr JELEŃ**[1], **Mirosław STYGAR**[1], **Juliusz DĄBROWA**[1], **Elżbieta DŁUGOŃ**[1], **Aleksander GIL**[1], **Krzysztof MROCZKA**[2], **Magdalena LEŚNIAK**[1], **Tomasz BRYLEWSKI**[1], **Maciej SITARZ**[1]

[1] *AGH University of Science and Technology, Faculty of Materials Science and Ceramics, Cracow, POLAND,*

[2] *Pedagogical University of Cracow, Institute of Technology, Poland*

e-mail: mbik@agh.edu.pl

In the 21st century, one of the most developing economic sectors are inseparable automotive industry and energy production. In order to provide with the highest efficiency of both of them, the greatest attention is drawn to a materials selection. Although, it is based on the best performance characteristics in working conditions, there occur many issues concerning corrosion resistance due to high aggressive environment combined of high temperature and harmful gaseous atmosphere. Two considered examples include ferritic steels for interconnects in Solid Oxide Fuel Cells and TiAl based alloys for turbine blades and combustion engines. Apart from promising ideas based on changing the chemical composition with additions (i.e. Nb or W leading to Laves phase formation) another approach is connected with surface modification in the form of protective coatings. As far as coatings are concerned, not only their high corrosion resistance is required, but also simplicity of production and low costs have to be taken into consideration.

In the following work there have been proposed coatings based on a brand new ceramic material of amorphous SiO_2 structure – black glasses. The beneficial local increase in density of bonds (mainly Si-C) resulting from the replacement of oxygen anions with carbon ones in a ratio equal to 2/1, leads to strengthening of a glass network. Because of that, coating material acquire a number of useful properties such as excellent thermal and chemical stability together with high oxidation resistance and mechanical strength. What is more, due to exceeding a given threshold of carbon content in a glass network, there can be observed a free carbon phase resulting in satisfactory electrical resistance, needed especially in Fuel Cells interconnects.

The aim of the following work was to evaluate the structure and microstructure of black glass coatings before and after oxidation on two kinds of substrate – ferritic stainless steel and intermetallic alloy.

The coating materials was obtained by the means of sol-gel method, where dip-coating technique was used to deposit a synthesized sol on previously grinded or polished samples. Modified alkoxysilanes were used as a ladder-like polysilsesquioxane gel precursor. Two-stage thermal treatment including drying (70°C, Air, 1 week) and pyrolysis (800°C, Ar, 0,5h) led to obtain black glass coatings. In order to confirm the formation of an appropriate material, a series of structural (MIR, XRD) tests were conducted. The key measurements for the preliminary evaluation of an oxidation resistance were thermogravimetric analyses in the laboratory air atmosphere in 800°C for 100h. A number of studies (XRD, SEM, EDS, Confocal) allowed to investigate microstructure and structure of a material before and after corrosive tests and to state the ability of black glasses to increase the corrosion protection of both of substrates.

This research was financed by Polish Ministry of Science and Higher Education from the budget for science in the years 2017-2020, as a research project under the program "Diamond Grant" (grant no. DI2016 004046). Also, Maciej Bik has been partly supported by the EU Project POWR.03.02.00-00-I004/16.

New carboxylate based corrosion inhibitors for mild steel.

Anthony SOMERS[1], Maria FORSYTH[1], Esther UDABE[2], David MECERREYES[2]

[1] *Institute for Frontier Materials, Deakin University, Australia*
[2] *POLYMAT, Spain*

Ionic inhibitors, such as rare earth carboxylates have shown promise as corrosion inhibitors for mild steel. Such compounds form a complex in solution and it has been suggested that the ability of one ion, in this case the carboxylate, to bring the counter ion, such as a rare-earth, to the surface results in improved performance compared to adding these components separately. Previous studies have shown that small changes in the structures of these carboxylate based corrosion inhibitors can make a large difference in their effectiveness to reduce corrosion. In this study a range of new carboxylate based inhibitor compounds have been assessed, with surface analysis used on selected samples to understand the difference in performance. These new compounds are based on coumaric acid combined with ammonium cations prepared to give 3 slightly different compounds. The first compounds were prepared as deep eutectic solvents (DESs), a relatively unexplored compound family as corrosion inhibitors. These materials contain a cation and hydrogen bond acceptor anion (eg chloride or acetate) that react with a hydrogen bond donor to form new materials with unique chemical properties. DESs can be inexpensive and simple to synthesise but still offer a diversity of possible chemistry. The second method was to prepare the compounds as protic ionic liquids to explore the effect of protonation on corrosion inhibition. Finally the compound was prepared as an ionic monomer, which could be incorporated into a polymer as part of a coating.

Lifetime Prediction of Yttria Stabilized Zirconia (YSZ), La2Zr2O7, Gd2Zr2O7, YSZ/ La2Zr2O7 and YSZ/ Gd2Zr2O7 Thermal Barrier Coatings (TBCs) with Cold Gas Dynamic Spray (CGDS) bond coat under Furnace Thermal Cyclic Loadings

Abdullah Cahit KARAOGLANLI[1], **Kadir Mert DOLEKER**[1]

[1] *Bartin University, Department of Metallurgical and Materials Engineering, Turkey*

Gas turbine engines under the service conditions were exposed to oxidation and thermal shock. Thermal barrier coatings (TBCs) are performed to protect metallic components of gas turbines against these degradations. Yttria Stabilized Zirconia (YSZ) is most common top coat materials for TBC systems. Nowadays, rare earth zirconates such as $La_2Zr_2O_7$ and $Gd_2Zr_2O_7$ are tried as an alternative to Yttria Stabilized Zirconia (YSZ) for TBCs. In this study, Inconel 718 substrates were coated with CoNiCrAlY using Cold Gas Dynamic Spray (CGDS) technique promising technology for bond coat spraying. All top coats were produced by Electron Beam Physical Vapor Deposition (EB-PVD) method providing high strain tolerance under thermal loadings. To predict the lifetime of TBCs, furnace cycling testing were performed to samples at 1150 °C. After thermal cycling tests, microstructural characterization of thermally grown oxide (TGO) layers, crack formations and chemical compatibilities between coating layers were investigated as detailed. Results show that double layered rare earth zirconate contented TBCs are higher lifetime compared to single layered TBCs.

Keywords: "Thermal barrier coatings (TBCs)", "cold gas dynamic spray (CGDS)", "yttria stabilized zirconia (YSZ)", "$La_2Zr_2O_7$", "$Gd_2Zr_2O_7$ "

Investigation of Microstructure, Oxidation, and Interface Properties of Single and Double-Layer Gadolinium Zirconate/Yttria Stabilized Zirconia Thermal Barrier Coatings (TBCs)

Abdullah Cahit KARAOGLANLI[1], Kadir Mert DOLEKER[1]

[1] *Bartin University, Department of Metallurgical and Materials Engineering, Turkey*

Oxidation is an inevitable failure mechanism under the operation temperature in gas turbines. To avoid negative effects of oxidation, ceramic based materials having low thermal conductivity and high stability should be used to hot section components. In accordance with this purpose, thermal barrier coatings (TBCs) are used in order to increase lifetime of gas turbine engine components which have not reached to desired levels yet. Yttria Stabilized Zirconia (YSZ) has been used as conventional top coat material in TBCs. Increased turbine inlet temperatures promotes to researchers to try higher stable material such as rare earth zirconates. In the present study, CoNiCrAlY metallic powders were sprayed using new emerging technique as called cold gas dynamic spray (CGDS) on Inconel 718 substrates. Single layer YSZ and double-layered YSZ/$Gd_2Zr_2O_7$ were deposited by Electron Beam Physical Vapor Deposition (EB-PVD) technique as top coat materials. In high temperature furnace, both TBC samples were isothermally oxidized at 1000 °C under different time periods. TBCs were examined as microstructural before and after oxidation tests. Thermally grown oxide (TGO) layer forming at the interface during oxidation were investigated and compared for each TBC systems. Oxidation and TGO growth behaviors were discussed.

Keywords: "Thermal barrier coatings (TBCs)"; "cold gas dynamic spray (CGDS)"; "oxidation"; "rare-earth oxides"; "$Gd_2Zr_2O_7$"

AUTHOR INDEX

AUTHOR INDEX

AUTHOR INDEX